Design for a Life

DESIGN FOR A LIFE

How Behaviour Develops

Patrick Bateson and Paul Martin

JONATHAN CAPE
LONDON

Published by Jonathan Cape 1999

2 4 6 8 10 9 7 5 3 1

First published in Great Britain in 1999 by
Jonathan Cape
Random House, 20 Vauxhall Bridge Road,
London SW1V 2SA

Random House Australia (Pty) Limited
20 Alfred Street, Milsons Point, Sydney,
New South Wales 2061, Australia

Random House New Zealand Limited
18 Poland Road, Glenfield,
Auckland 10, New Zealand

Random House South Africa (Pty) Limited
Endulini, 5A Jubilee Road, Parktown 2193, South Africa

Random House UK Limited Reg. No. 954009

A CIP catalogue record for this book
is available from the British Library

ISBN 0–224–05064–8

Papers used by Random House UK Limited are natural, recyclable products made from wood grown in sustainable forests. The manufacturing processes conform to the environmental regulations of the country of origin.

Typeset by Deltatype Ltd, Birkenhead, Merseyside
Printed and bound in Great Britain by
Biddles Ltd, Guildford and King's Lynn

Contents

I

The Developmental Kitchen

What is the use of a new-born child?
Benjamin Franklin

From Egg to Adult

How and why does each human grow up to be a unique individual? What role do genes play in shaping behaviour and personality? Are people's characters fixed early in life or can they change as adults? How does early experience affect sexual preferences? Why do children play? These are all questions about behavioural development – the lifelong process of growth and change from conception to death that is central to an understanding of human nature.

After the microscope was invented in the late sixteenth century, people gazed with excitement at the structure of the fertilised egg and thought they could see dimly within it the makings of an adult. Some even saw a small person crouched inside the head of each human sperm – or, if their prejudices were different, inside the unfertilised egg. It seemed satisfying to think of growing up as merely getting larger. But the satisfaction was deeply misplaced. Most animals, including humans, do not just grow – they develop. Everybody is the product of development.

The fertilised egg from which each person develops is barely visible to the eye. The adult human body consists of millions upon million of cells doing different things. Nerve cells, skin cells and white blood cells, for example, all contain the same set of genes but are highly specialised for particular tasks. The cells are integrated within organs devoted to specific functions, such as digesting food, pumping nutrients and oxygen to remote corners of the body, and integrating body movement in ways that serve the overall need of the whole individual to survive and reproduce. How does an organism such as a human transform itself from a tiny single cell into a self-aware individual with organs and cells all arranged in the right place at the right time?

Each human is a product of essentially the same developmental processes. And yet each human is also a distinctive individual. As Friedrich Nietzsche wrote, 'At bottom every man knows well enough that he is a unique being, only once on this earth; and by no extraordinary chance will such a marvellously picturesque piece of diversity in unity as he is, ever be put together a second time.' Understanding behavioural development means understanding the biological and psychological processes that build a unique adult from a fertilised egg. It also means understanding how these developmental processes have themselves evolved, and how their features can be viewed in terms of their biological design properties. But it does *not* mean trying to explain human behaviour in terms of the conventional opposition between nature (genes) and nurture (environment).

'Nature' and 'Nurture'

In his autobiography, Charles Darwin expressed this view about the development of human mental faculties, 'I am inclined to agree . . . that education and environment produce only a small effect on the mind of any one, and that most of our qualities are innate.'[1]

A contrasting view of how each person's mind is formed was expressed by the seventeenth-century philosopher John Locke:

> Let us then suppose the mind to be, as we say, white paper, void of all characters, without any ideas; how comes it to be furnished? . . . Whence has it all the materials of reason and knowledge? . . . To this I answer, in one word, from experience; in all that our knowledge is founded and from that it ultimately derives itself.

Many people end up believing in a bit of both – some qualities reflect 'nature' and some 'nurture'. Attitudes to whether a skill reflects an unchangeable part of the child, or the way the child has been taught, may depend on how essential that skill is thought to be. Parents who would baulk at the suggestion that their children's personalities and behaviour are 'all in the genes' can simultaneously be attracted by the notion that more exotic talents, such as exceptional abilities in music or mathematics, are inherited or 'innate'. Ambitious parents who incline to this view may feel less compelled to pressurise a child who displays no obvious flair for the subject. After all, the argument goes, you've either got it or you haven't, and if you haven't then no amount of hard work or practice can help. The selfsame parents will usually adopt a quite different view of development if their children are found to be deficient in a skill such as reading. Now the focus will switch to environmental influences. Their children are useless at music because music does not run in the family, but they cannot read well because their teachers are incompetent.

Attitudes towards development also change greatly with fashion and with shifts in ideology. For many years after the Second World War the mere suggestion that genes influence human behaviour was regarded as distasteful – largely, and understandably, in reaction to the appalling abuses of science that culminated in genocide by the Nazis. Abuses were also found on the opposite side of the ideological divide. Two years after the Russian Revolution, Lenin is said to have paid a secret visit to the laboratory of the great physiologist Ivan Petrovich Pavlov to find out if Pavlov could help the Bolsheviks to control human behaviour.[2] Lenin believed that individualism would interfere with his plans for a Communist state. Pavlov told him that 'natural instinct' could be abolished by conditioning – a form of learning. So congenial was this opinion to Lenin that Pavlov's position in the Soviet Union was thereafter protected, despite his not being a Communist. Long after Pavlov's death a big government-sponsored scientific congress was held in Moscow in 1950 devoted to his teaching. The party line was that humans could transcend heredity and be controlled by education. A second prong in this Stalinist programme enforced as dogma the view that all acquired characteristics were passed on to subsequent generations. Under the advocacy of this dogma by T.D. Lysenko in the late 1940s, Soviet genetics effectively ceased to exist.

One cynic commented, 'The environmentalists seem to believe that if cats gave birth in a stove, the result would be biscuits.' Since then, the pendulum has swung a long way in the opposite direction, and nowadays it sometimes seems as if almost any aspect of human behaviour or physiology can be accounted for in terms of genes alone. With increasing frequency the media report the discovery of a gene 'for' some distinct human characteristic, such as learning foreign languages, athletic prowess or male promiscuity.

It is obvious that experience, education and culture make a big difference to how people behave, whatever their genetic inheritance. Yet behavioural and psychological development are frequently explained in terms of the exclusive importance of one set of factors, either genetic or environmental. Such firmly held opinions partly derive from a style of advocacy common to most scientific debates. If Dr Jones has overstated her case, then Professor Smith feels bound to redress the balance by overstating the counter-argument. The confusions are amplified because of the way in which scientists analyse developmental processes. When somebody has conducted a clever experiment demonstrating an important long-term influence on behaviour, they have good reason to feel pleased. It is easy to forget about all those other influences that they had contrived to keep constant or which play no systematic role. Consequently, debates about behavioural and psychological development often degenerate into sweeping assertions about the overriding importance of genes (standing in for 'nature') or the crucial significance of the environment (which then becomes 'nurture').

Design for a Life

The various stages of an individual's development, from conception, through childhood, adolescence, parenthood (for most) and the later years of maturity and senescence, are completely different, and the individual must function at least adequately during each of them. Even though it is easy to think of exceptions where brain and behaviour malfunction terribly, as in paranoid schizophrenia, much of what happens during the process of development looks as though it is well designed. The proposition that living organisms' bodies, brains and behaviour were adapted over the course of evolution to the conditions in which they lived is at least familiar to most non-biologists. An adaptation is a modification that makes the organism better suited to survive and reproduce in a particular environment – better suited, that is, than if it lacked the crucial feature.

In the early part of the nineteenth century, the English theologian William Paley wrote incisively about the adaptations of biology. In his book *Natural Theology*, Paley emphasised how different parts of the body relate to each other and contribute to the whole. He illustrated this by considering the various features of the mole:

> The strong short legs of that animal, the palmated feet armed with sharp nails, the pig-like nose, the teeth, the velvet coat, the small external ear, the sagacious smell, the sunk protected eye, all conduce to the utilities or to the safety of its underground life . . . The mole did not want to look about it; nor would a large advanced eye have been easily defended from the annoyance to which the life of the animal must constantly expose it. How indeed was the mole, working its way under ground, to guard its eyes at all? In order to meet this difficulty, the eyes are made scarcely larger than the head of a corking-pin; and these minute globules are sunk so deeply in the skull, and lie so sheltered within the velvet of its covering, as that any contraction of what may be called the eye-brows, not only closes up the apertures which lead to the eyes, but presents a cushion, as it were, to any sharp or protruding substance which might push against them. This aperture, even in its ordinary state, is like a pin-hole in a piece of velvet, scarcely pervious to loose particles of earth.
>
> Observe then, in this structure, that which we call relation. There is no natural connection between a small sunk eye and shovel palmated foot. Palmated feet might have been joined with goggle eyes; or small eyes might have been joined with feet of any other form. What was it therefore which brought them together in the mole? That which brought together the barrel, the chain, and the cogs, in a watch – design; and design, in both cases, inferred from the relation which the parts bear to one another in the prosecution of a common purpose . . . In a word; the feet of the mole are made for digging; the neck, nose, eyes, ears, and skin are peculiarly adapted to an underground life; and this is what I call relation.

Paley, who became a bishop, regarded the design he saw everywhere in nature as proof of the existence of God. These days few biologists would try to pin their religious faith on biological evidence, and the design to which Paley referred would be attributed instead to the evolutionary mechanism which Charles Darwin called natural selection. Darwin's theory of evolution by natural selection is universally accepted among scientists, even if arguments continue over the details. Darwin proposed a three-stage cycle that starts with random variation in the form and behaviour of individuals. In any given set of environmental conditions some individuals are better able to survive and reproduce than others because of their distinctive characteristics. The historical process of becoming adapted notches forward a step if the factors that gave rise to those distinctive characteristics are inherited in the course of reproduction. Suppose, for example, that an individual bacterium happens to have heritable characteristics that make it resistant to the latest antibiotic. While all the others are killed by antibiotics, this one will survive and multiply rapidly. Before long, the world (or, at least, the hospital ward) is full of antibiotic-resistant bacteria. Darwinian evolution requires no unconscious motives for propagation – let alone conscious ones. In Richard Dawkins's phrase, 'The world became full of organisms that have what it takes to become ancestors.'

Biologists have been properly warned not to write evolutionary accounts in which the past is seen as leading purposefully towards the goal of the present blissful state of perfection. A clear distinction is necessarily and wisely drawn between the present-day utility (or function) of a biological process, structure or behaviour pattern, and its historical, evolutionary origins. Darwin noted, for example, that while the bony plates of the mammalian skull allow the young mammal an easier passage through the mother's birth canal, these same plates are also present in the mammals' egg-laying reptilian ancestors. Their original biological function clearly must have been different from their current one.[3]

The distinction between current function and historical evolution is all the more necessary because current adaptations may result from the experience of the individual during its lifetime. Human hands form calluses to protect against mechanical wear, and muscles develop in response to the specific loads placed upon them during exercise. Behaviour, in particular, becomes adapted to local conditions during the course of an individual's development, whether through learning by trial and error or through copying others. These are all examples of adaptations that are acquired during the lifetime of the individual, and they are clearly distinct from adaptations that are inherited. 'Design' in this book is used in its widest sense to denote adaptations, whatever their biological origins.

The Developmental Menu

A young man who was contemplating a career as a chef visited a country hotel which prided itself on using only the best locally grown produce. He was solemnly told that the only thing that matters in good cooking is to ensure the ingredients are of high quality. The young man was not stupid, however, and felt this advice must be incomplete. Then he consulted the celebrated chef at a fashionable urban restaurant and was told emphatically that what really matters is having the right kitchen equipment and paying meticulous attention to the presentation of the food at the table. 'You can make anything look beautiful and taste marvellous.' Our novice chef sensed that both his advisers had been claiming too much. Their prejudices were compounded by self-righteousness when they heard about the views of the other. Though as yet unversed in the mysteries of cooking, the young man realised intuitively that the ingredients of a meal and the way in which they are put together must both matter. He went on to become a much better chef than either of his advisers.

The processes involved in behavioural and psychological development have certain metaphorical similarities to cooking. Both the raw ingredients and the manner in which they are combined are important. Timing also matters. In the cooking analogy, the raw ingredients represent the many genetic and environmental influences, while cooking represents the biological and psychological processes of development. Nobody expects to find all the separate ingredients represented as discrete, identifiable components in a soufflé. Similarly, nobody should expect to find a simple correspondence between a particular gene (or a particular experience) and particular aspects of an individual's behaviour or personality. This point, which we hope will be obvious, is central to much of what we describe in this book.

'Man is born to live, not to prepare for life,' wrote Boris Pasternak in *Doctor Zhivago*. The behaviour seen at a particular stage of life may be part of the cooking process of development, but it may also serve a current need. Many attributes of a young, developing organism are adaptations to the environment it is inhabiting at that stage in its life. In order to become an adult, the young animal or child must obviously survive in the here and now. But some of what goes on in development is also about preparing for the future, setting the menu for the life yet to come. This forward-looking aspect of development was expressed by Simone de Beauvoir in her autobiography:

> All through my childhood and my young days, my life had a distinct meaning: its goal and its motive was to reach the adult age. At twenty, living does not mean getting ready to be forty. Yet for my people and for me, my duty as a child and an adolescent consisted of forming the woman I was to be tomorrow.

Undoubtedly, many important aspects of behavioural and

psychological development do relate to the individual's future rather than current needs. Certain adaptations – such as play behaviour – assist the developmental process of acquiring skills or knowledge of the social and physical environment which will be needed in later life. Preparation for the future may make use of predictive information about the environment which the individual will inhabit. The need to predict becomes especially important when the physical or social environment may take radically different forms, forcing individuals to use quite different solutions according to the particular problems they face. The habitats in which humans may end up are dramatically different from each other; some are demanding physically and others are demanding socially. Individuals are prepared in some measure during their early development for the particular adult world they will enter.

Inevitably, much of what we shall describe in this book reflects common experience. We shall draw from time to time on the writings of novelists and poets who have expressed themselves more memorably, amusingly or imaginatively than we could ever have done. We hope that this interweaving of science and literature will seem as natural to others as it did to us. When dealing with substance rather than method, we believe that the wall built between the sciences and the humanities is artificial and unhelpful – and nowhere more so than in the kitchen of human development.

2

The Seven Ages

That which we call 'development' when looked at from the birth end of life becomes senescence when looked at from its close.

Peter Medawar, *The Uniqueness of the Individual* (1957)

Players on the Stage

Every individual must act many roles during his or her lifetime. Through Jacques in *As You Like It* Shakespeare drew attention to these very different parts as each person grows up and ages:

> All the world's a stage,
> And all the men and women merely players:
> They have their exits and their entrances;
> And one man in his time plays many parts,
> His acts being seven ages.

Shakespeare's choice of seven periods is not sacrosanct. John Burrow describes in his book *The Ages of Man* how a number of different ways of dividing up life had entered medieval thought from antiquity.[1] Aristotle had distinguished the period of growth, the prime of life and the period of decay. The Venerable Bede had four ages: childhood, which is metaphorically moist and hot;

youth, which is dry and hot; maturity, which is dry and cold; and old age, which is cold and moist. The seven ages used by Shakespeare, and before him the medieval astrologers, came originally from the second-century astronomer Ptolemaeus and were based on the main heavenly bodies. In the first four years of life the moon produces change. In the following ten years Mercury articulates and fashions the logical mind. In the third age Venus implants the impulse for love. From the ages of twenty-two to forty-one the sun provides mastery and direction, and changes playfulness to seriousness and ambition. From forty-one to fifty-six Mars brings unhappiness and a desire to accomplish something before the end. Jupiter brings thoughtfulness and dignity to the sixth age. The last age, which belongs to Saturn and starts at the age of sixty-eight, is one of cooling and slowing down.

Those who study psychological development in modern times argue among themselves about whether development is continuous or discontinuous. If they focus on continuities between one age and the next, it does not make much sense to split development up into discrete stages. If they focus on the sudden changes, then these provide natural boundaries. In later chapters we shall point out the continuities in development, but here we consider the steps and what they might mean.

Why, for instance, is the infant, 'mewling and puking in the nurse's arms', so different from the 'whining school-boy, with his satchel and shining morning face'? Do the profound changes that occur during an individual's lifespan represent, like Shakespeare's seven ages, different roles that must be acted in successive scenes? The job of scientists who study the development of behaviour is to make sense of the seven scenes – or, rather, eight scenes, since the phase before birth must also be considered. Life starts at conception and a great deal happens during the nine months before birth.

An animal at the start of its lifespan is not just a miniature adult which simply has to grow in size. It would be badly designed if it were. Indeed, a developing animal might as well be another species, so different is it from what it will become in adulthood. At the start of human development the fertilised egg is barely a tenth of a millimetre in diameter. It does not even have a brain. Twenty-five years later the same individual could be composing a symphony or flying an airliner.

The profound transformations that occur during the course of an individual's lifespan are dramatically illustrated by the metamorphosis of a caterpillar into a butterfly, or a tadpole into a frog. But something akin to metamorphosis goes on in human development as well. Before birth the human foetus is surrounded by liquid and its oxygen comes directly from its mother's bloodstream. After birth the baby breathes air and is launched into a terrestrial environment. Subsequent transitions may not be quite so dramatic, but the onset of sexual maturity brings a new body and all the preoccupations and cares of adulthood.

The environment of the developing human changes radically as it grows. The individual starts life inside the mother's womb, is born, gradually becomes independent and eventually may become a parent itself. At each successive stage in its development, the individual must have special ways of meeting its changing requirements. Many aspects of its behaviour and physiology are designed to meet needs that arise from that current stage of development. Once that stage has passed, these specialised behaviour patterns may drop out, like milk teeth, never to be used again. Consider, for example, the different ways in which people feed themselves during the course of their lifespans. The foetus acquires nourishment passively, through the placenta; the baby by suckling from its mother's breast; and the adult by seeking out and eating solid food. Or consider the time that people allocate to reproduction; before puberty no time at all and afterwards maybe a lot. The child simulates what the adult does but with little cost, playing with toys, whereas the adult copes with the babies, tools and weapons on which the toys are modelled. Much of what the child does is necessary for construction of the adult it will become, but not required when the adult design is complete. So, for example, young children are equipped with special mechanisms enabling them to acquire language.

The age of puberty is influenced by environmental conditions, but humans are ready to breed in their second decade. By the beginning of their sixth decade, they are much more likely to be nurturing their grandchildren than having their own children. These broad characteristics of a species' lifespan are the products of conflicting evolutionary pressures. They are the result of a trade-off roughly equivalent to that of a gambler deciding between staking a large amount of money on one bet or making many small bets. How many offspring are conceived, and how much parental care is invested in each one, will depend on the resources available to members of that species, how big the individuals are as adults, how likely they are to survive from one year to the next, how many skills they need as a successful adult, and much else.[2]

Most adults of most species are able to produce more than one offspring during their lifetime. The characteristics of those offspring that are most successful in leaving descendants will tend to predominate in subsequent generations. Parents who sacrifice too much for one offspring will have fewer descendants. By the same token, offspring who do less to ensure their own survival than others will have fewer descendants. Such are the broad rules that shape any lifespan, but just how they look in detail will depend on the species and, within a species, on local conditions.

Parents, Offspring and Conflict

In some species, parental care for the tiny progeny consists of nothing more than providing a small amount of yolk, enough to sustain the offspring until they can feed for themselves. Marine fish such as herring, for instance, produce vast numbers of eggs and sperm which fuse in the sea. Neither parent provides any care for their progeny. Most fertilised herring eggs consequently die at an early stage in their lives. Other fish produce far fewer fertilised eggs and care for them in a variety of different ways, some keeping them like a mammal in their bodies until the young are born, others gathering the fertilised eggs into their mouths, where they are protected.

Of all animal groups, birds and mammals produce the smallest number of young and take the greatest care of those they do produce. Birds encase their fertilised eggs in a hard shell and eject them from their bodies, whereupon both sexes usually take turns in keeping the developing egg at body temperature until the chick hatches. While it is developing inside the egg the embryonic chick is fed from the yolk, which at the outset is enormous relative to the embryo. After the egg has hatched, both parents usually protect and bring food to the developing young.

Humans, like many mammals, invest heavily in each of a small number of offspring during the course of their reproductive lifespan. This has important consequences for the amount of care that is given by the parent and the time taken to become adult. The long period of development that is particularly characteristic of humans, but also true to a lesser extent of most other mammals, is made possible by the protection and the provisioning by the parent. Humans take a long time, even relative to their long lifespan, to reach reproductive age. As they prepare for their own eventual reproductive life, children meanwhile have to survive.

The young mammal is well protected inside its mother's body for the first part of its life. When the early embryo attaches itself to the lining of the womb, which happens six to eight days after fertilisation in humans, cells derived from the offspring grow like a parasite, penetrating the membrane that lines the womb and altering its blood supply. Arteries in the womb lining are turned into low-resistance blood vessels which cannot constrict. The offspring's side of this remarkable connecting device is the placenta. The unborn offspring is entirely dependent upon it. The placenta functions in place of a mouth, lungs and kidneys. While the blood of mother and offspring never come in contact, food and oxygen from the mother's bloodstream are passed across the wall of the womb and placenta to the offspring's own blood, while its waste products pass the other way. Once the exchanges have occurred, the offspring's blood passes back through the umbilical cord to the growing offspring. After birth, the umbilical cord is severed and the placenta is shed from the mother's body as the afterbirth.

The traditional image of parenthood has been one of complete harmony between the mother and her unborn child. But evolutionary theory has cast doubt on this blissful picture. In sexually reproducing species, parents are not genetically identical to their offspring. Consequently, offspring may require more from parents than parents are prepared to give, creating the possibility of a conflict of long-term interests. The American biologist Robert Trivers called this 'parent–offspring conflict', a term which refers strictly to a conflict of reproductive interests, not conflict in the sense of overt squabbling.[3] The parent may sacrifice some of the needs of its current offspring for others which it has yet to produce; the offspring maximises its own chances of survival. Such conflicting interests need not – and usually do not – involve conscious thought on the part of parent or offspring. Moreover, the potential for conflict starts long before the offspring is born. Consider what goes on in the womb during pregnancy.

The placenta invades the mother's body in ways that are not always to her long-term benefit in terms of having other children. The Australian biologist David Haig has suggested that while it is in the offspring's interest to maximise the transfer of nutrients across the placenta, the mother's best interest lies in limiting the transfer of nutrients to a somewhat lower level.[4] Mother and offspring 'disagree' about how much the offspring should receive. Haig argued that placental hormones that pass into the mother's bloodstream manipulate her physiology to the benefit of the foetus. By raising the mother's blood pressure, the flow of blood to the placenta is increased, thereby increasing the supply of nutrients to the foetus. Usually when this happens the human mother merely suffers swollen ankles. But occasionally a pregnant woman develops eclampsia – a potentially fatal condition – and she may suffer convulsions and lapse into a coma. Haig argued that placental hormones, derived from the offspring, inhibit the effects of insulin, thereby maintaining blood-sugar levels at a higher level for longer periods after she has eaten than when she is not pregnant. Sometimes she is pushed too far and may end up suffering from diabetes.[4]

The result of such evolutionary conflicts of interest is sometimes portrayed as a form of arms race, with escalating foetal manipulation of the mother being opposed by ever more sophisticated maternal counter-measures. However, limits must be encountered in the course of evolution. If the offspring is too aggressive in its demands it will kill its maternal host and, of course, itself. Likewise, if the mother is too mean, her parasitic offspring will not thrive and she might as well have not bred.

Placentas are strikingly different in structure from one mammalian group to the next. Generally, the foetus constructs three layers of tissue in making the attachment to the mother's womb. Depending on the species to which she belongs, the mother may have anything from three layers on her side of the placenta to none, in which case the maternal blood freely bathes the foetal intrusion into her womb. In yet another type of placenta, the maternal blood freely bathes only one layer of foetal membrane.[5]

Attempts to classify species of mammals according to the number of layers between the bloodstreams of mother and foetus do not correspond to any of the other classifications of mammals that have been constructed on the basis of similarities of bodily structures or genes. So, for example, horses, whales and lemurs (which are primitive primates) all have three maternal and three foetal layers of placenta separating their bloodstreams. Shrews, rats and rabbits, on the other hand, are similar to the main group of primates, including humans, in having free maternal blood bathing a single foetal layer. Nobody quite knows why this should be. The precise number of layers that has evolved is likely to have depended on how advantageous it was for both parties that the offspring should influence the mother's body: the fewer the layers, the greater the scope for influence. The number of placental layers may have flipped backwards and forwards during the evolution of the species, as the optimum balance of interests between parents and offspring shifted. The evolutionary history of a species cannot be deduced from the current structure of its placenta.

The different interests of parent and offspring are described in Kahlil Gibran's poem *The Prophet*:

> Your children are not your children.
> They are the sons and daughters of Life's longing for itself.
> They came through you but not from you
> And though they are with you yet they belong not to you . . .
> You are the bows from which your children as living arrows are sent forth.

Life in the Womb

Premature babies are tiny but completely formed as much as three months before the normal time for birth. Their tactile, olfactory, auditory and visual systems have been established, in that order.[6] The main nervous connections to the frontal part of the brain are intact. Any mother knows how active her unborn child can be, especially in the final three months of pregnancy. Scans with ultrasound reveal a remarkable picture of unborn children engaged in a variety of activities, including stretching, kicking and sucking their thumbs. We shall discuss in later chapters how much the foetus is affected by the chemical environment provided by its mother, and how it may prepare itself for the external environment that is predicted by its mother's state.

The notion that individuals are affected by what happens to them in the womb – in other words, that behavioural development starts before birth – is an old one. John Donne wrote this forbidding passage in 1630: 'In the womb we have eyes and see not, ears and hear not. There in the womb we are fitted for works of darkness, all the while deprived of light. And there in the womb we are taught cruelty, by being fed with blood, and may be damned though we be never born.'

Donne was fanciful about cruelty being taught before birth, but much *is* experienced by the foetus. It is exposed to much more sensory stimulation than was once supposed and its sentient life, especially in the last three months before birth, is now well recognised. It can, for example, hear the sound of its mother's voice in the womb. So, with a functional auditory system, the foetus will hear her and, as is now known, will start to learn what her voice sounds like.

Newborn babies show a distinct preference for their mother's voice, over and above the sounds of other women's voices.[7] In one experiment, an ingenious method was used to find out whether this preference for mother's voice was affected by experience before birth.

Each newborn baby in the study was given a dry teat to suck. Different babies habitually sucked their teats at different, individually distinctive, rates. When the characteristic sucking rate for each baby had been established, half the babies were played a recording of their mother's voice – but only if they sucked at a lower rate than was usual for them. Otherwise, the baby heard a recording of an unfamiliar woman's voice. The procedure was reversed for the rest of the babies, so that they had to suck at a higher rate than usual to hear their mother's voice. These newborn babies soon learned to modify their normal sucking rate to one that was rewarded with the sound of their mother's voice. They demonstrated not only that they preferred to hear their mother's voice, but also that they were capable of learning how to modify their behaviour in order to bring this about.[8]

A second experiment produced even more remarkable results. During pregnancy ten mothers regularly read aloud, as if to their unborn child, a story called 'To think that I saw it on Mulberry Street'. Later, after they had been born, these same babies were given the opportunity to suck, either more or less frequently than usual, in order to hear a recording of their mother reading 'To think that I saw it on Mulberry Street', as opposed to another story. The newborn babies learned to alter their sucking behaviour in order to hear not just their own mother's voice, but also the sound patterns of the particular story that had been repeatedly read to them before they were born. Their post-natal behaviour and their preferences had clearly been influenced in a quite specific way by their experiences in the womb.

The learning by unborn babies of their mother's voice and also her smell prepare them for the process of social attachment that takes place after birth. While other cues provided by the mother may set the baby on a particular trajectory of subsequent development, much of what happens in the womb is unique to this stage of the life cycle and crucial to the process of rapid growth. Meanwhile, the foetus is protected from cold, want and disease as it never will be again.

A Brave New World

Birth is often held to be traumatic for the baby. Some remarkably unconvincing memories of travelling down the birth canal have supposedly been recovered from people in later life. These probably have much more to do with adult projection than reality.[9] Even so, the departure of the baby from its uterine environment is sudden and sometimes damaging if the birth process is prolonged. The baby may be injured by the attempts to extricate it from its mother's womb. Or it may suffer from anoxia if the supply of oxygen from the mother is disrupted and it is unable to draw its first breath.

In the first few days after birth, much happens to secure a healthy and successful passage to the next stage. The mammalian mother produces a specialised food – milk – from an external organ which the offspring sucks periodically. Apart from nutrition, the initial surge of mother's milk, the colostrum, helps to immunise the baby against bacteria which it is bound to ingest now it is exposed to the outside world. Suckling by the offspring is another adaptation to a particular stage of life which falls away later in life. The method of taking in milk and the motivation to do so are controlled in different ways from adult feeding and drinking. A few days after birth rat pups may be induced to lap milk like an adult.[10] This is at an age when they are still suckling. The adult feeding pattern becomes progressively more vigorous the longer the pup is deprived of food, whereas the suckling is as vigorous immediately after the pup has received milk as it is hours later.

In the hours and days after birth a strong emotional bond is usually formed between mother and baby. If the mother is unable for medical reasons to have immediate contact with her baby, the bonding process may be slower and more troublesome.[11] The window within which the bond forms has sometimes been over-dramatised, in that some mothers have been led to believe that the bonding will be impossible without immediate contact with their baby. That is ill-founded, since mothers who have been separated from their babies are able to form strong attachments to them later.[12]

'Begin, baby boy, to recognise your mother with a smile,' wrote the Roman poet Virgil more than two thousand years ago. Mothers and their babies gaze at each other and the interplay between them becomes as well timed as dancing. Lynn Murray and Colwyn Trevarthen discovered how much the timing matters. When mother and child made contact through television cameras and screens, the relationship was maintained – until an artificial delay was introduced. If the baby saw what the mother had done a few seconds before in response to its own actions, the timing was upset and so was the baby.[13] Away from the psychologist's laboratory, the exquisite timing of their shared activities brings great satisfaction to both mother and baby, maintaining a strong bond between them.

In the 1970s, psychologists documented how the baby would copy the actions of the mother. If she protruded her tongue, so would the baby; if she formed her lips into an O, so would the baby.[14] One hundred years before this, Charles Darwin described what he thought was imitation of sounds by his four-month-old son.[15] Darwin's careful observations, which we shall refer to again, are often regarded as the foundation of the study of behavioural development. Darwin observed, as many have done since, the pleasurable and playful character of the baby's relationship with both parents. Smiling to a parental face, with the eyes brightening and the eyelids closing slightly, began at forty-five days. When Darwin made strange noises, his infant son regarded these as 'excellent jokes'. As many other parents have discovered, hiding and then uncovering his own face caused great amusement in his son.

Even at birth, babies notice and store much more about their world than was once attributed to them. Compared to other mammals, though, humans are slow to move their bodies with any effectiveness. Little by little the baby acquires motor skills. The median age for transferring an object from one hand to the other is five and a half months and a neat pincer grasp has developed by nine months in most human infants.[16] The range of variation is great, but more than half are standing unaided by eleven months and walking by a year. Infants absorb a great deal of information and come to terms with the structure and dynamics of the world about them – assimilating and accommodating, in the phrase of the Swiss psychologist Jean Piaget.

Like Darwin, Piaget watched his own children carefully. He advanced a famous theory of how children develop their capacities to understand and control their environment. Piaget believed that development was step-like, with quite distinct shifts in ability.[17] Like changing gear as a car gathers speed, the skills of assimilating new information and accommodating to what has been discovered change qualitatively with age. The precise ages of these gear changes are disputed. By degrees, they have been placed ever earlier in development than Piaget would have had them, as developmental psychologists have invented increasingly ingenious techniques for probing the abilities of babies and infants in ways that do not depend on understanding spoken questions.

The Beginning of the End

Weaning – the process of changing from taking milk to eating solid food – may be gradual or sudden, depending on the species. The young of plant-eating species start to forage for themselves shortly after birth, although they cannot immediately survive without their mother's milk. The young of carnivorous mammals have their first meal of meat weeks after they are born, when one or other of the parents brings a kill back to the den. Human infants generally do not take any solid food until they are at least three to four months old.

The evolutionary concept of parent–offspring conflict, so

interesting when considering the evolution of the placenta, is less helpful when applied to the process of weaning. It led to a false expectation that a conflict of evolutionary interests necessarily implies observable squabbling between parents and offspring. Direct observations of many mammals show that maternal aggression towards offspring rarely occurs and, if it does, is not seen at weaning, the stage at which evolutionary theory predicted it was most likely to occur.[18]

By the age of two, most young humans are fully weaned, even though they may remain dependent on one or both parents for many years. Two is the beginning of the end, suggested J.M. Barrie in *Peter Pan:*

> All children, except one, grow up. They soon know that they will grow up, and the way Wendy knew was this. One day when she was two years old she was playing in a garden, and she plucked another flower and ran with it to her mother. I suppose she must have looked rather delightful, for Mrs Darling put her hand to her heart and cried, 'Oh, why can't you remain like this for ever!' This was all that passed between them on the subject, but henceforth Wendy knew that she must grow up. You always know after you are two. Two is the beginning of the end.

Tantrums in two-year-old children ('the terrible twos') seem to be concerned with the difficult process of establishing autonomy from the parent, rather than with the conflicts of interest between parent and offspring. Children who are emotionally well attached to their parents rarely have such tantrums, but instead use their parents as a secure base from which to explore the world.[19] The child tests its parents as it unconsciously searches for the boundaries of what is acceptable behaviour. Wise parents will understand what is happening and know that even their genuine anger at something particularly outrageous or infuriating on the child's part conveys useful information to the child.

The nature of the relationship between siblings is somewhat more in accord with evolutionary theories about conflict than is the parent–offspring relationship. As in the case of parents and offspring, the biological interests of siblings are not identical.[20] Given that parental resources are finite, anything given to a sibling is a potential loss to yourself. Some striking examples of overt conflict between siblings are found in the animal kingdom. Many species of birds such as the herons, egrets and pelicans hatch out two or three young at intervals of several days. If food is plentiful all of them survive. But if food is scarce, the youngest offspring is killed by its older siblings, with no attempt by the parents to interfere. Indeed, the parent birds may feed their dead offspring to the survivors.[21]

Human siblings are rarely as brutal in their treatment of each other. Conflicts of interest may nonetheless manifest themselves in more subtle forms. For instance, under conditions where food is scarce, a child's growth rate during the weaning period is reduced if its mother becomes pregnant again.[22] The older child's milk supply is curtailed so that its unborn sibling can survive. Charles Darwin noted that his fifteen-month-old son reacted with jealousy when he fondled a large doll or when he weighed his son's infant sister: 'Seeing how strong a feeling jealousy is in dogs, it would probably be exhibited by infants at an earlier age.' The unfairness of the way he was treated relative to his older sister Fanny was keenly felt by Charles Dickens and was reflected in his thinly disguised portrait of his own childhood, *David Copperfield*. Dickens was set to work in a factory at twelve, whereas his sister Fanny was sent to the Royal Academy of Music.

Such aspects of sibling rivalry have been carefully documented by the developmental psychologist Judy Dunn and her colleagues.[23] They showed that children are highly sensitive to the way that parents respond to their siblings. The arrival of a new baby has a marked effect on the behaviour of the first child, even when it is as young as fourteen months. The first-born usually demands greater attention. Later-born children, as they enter their second year, carefully monitor their mother's relationships with older siblings, minding a great deal when they receive less attention.

They Tuck You Up

The long period from weaning to total independence is a striking feature of human development, raising fundamental questions about the adaptive significance of such a distinctive aspect of human biology. A group of other features is associated with this prolonged development, including repeated single births, intense parental care and membership of highly social groups. A prolonged period for learning, free from the cares of finding a mate and caring for young, greatly facilitates the acquisition of the detailed knowledge and complex skills that are required in transactions with equally well-informed and skilled members of the species. Since a slip backwards to a shorter period of growth would put those individuals at a grave disadvantage, the evolutionary pressure has been to ratchet up those capacities that are required in both competitive and co-operative aspects of social life.

While a long period of development is clearly valuable, human flexibility is such that the time for building a detailed bank of knowledge and skills can be sharply curtailed if this becomes necessary. In Chapter 6 we shall describe the astonishingly resourceful street children found in many big cities of the world. When their survival depends on it, children can cope as though they were adults. The longer-term costs to them may be great, but it is better to pay those costs than not survive at all.

As children grow older, become mobile and start to communicate and play, they form peer groups with their own particular ways of behaving towards each other. Judith Harris has argued that these peer groups provide the dominant influence on children – despite most parents' belief that they are the ones who shape their child's character.[24] Children tend to pick up their attitudes and accent from their peers, not from their parents or teachers. Peers are important. But parents play their part even here because they are able to have a strong influence, whether wittingly or unwittingly, on who those peers are.

Home provides only one place in which children must learn how to behave. They may behave differently in other environments – for example, at school – and find it difficult when their parents visit the school or school friends visit their home. Children are resistant to parental attempts to raise boys and girls in the same way; they divide themselves into single-sex peer groups, where they expose each other to behaviour typical of that sex, despite the efforts of well-intentioned parents to avoid gender stereotypes.

Childhood is viewed romantically as a trouble-free period by some and a time of conflict by others. George Bernard Shaw gave a characteristically cynical twist: 'Youth is a wonderful thing; what a crime to waste it on children.' The darker side of childhood was described by Samuel Butler in his semi-autobiographical novel *The Way of All Flesh*, in which he gloriously debunked the nineteenth-century idealisation of childhood and the parent–offspring relationship. Although it was not published until 1903, after Butler's death, *The Way of All Flesh* was written between 1872 and 1884, when Darwin's *Origin of Species* was being much debated. Butler himself was a sharp and public critic of Darwin, believing that Darwin had slighted the earlier writers about evolution, particularly his grandfather Erasmus Darwin. Like Ernest Pontifex, the central character of his novel, Samuel Butler was raised in a rectory by a stern and unbending clergyman father. *The Way of All Flesh* tells the story of a son rebelling against parental control and hypocrisy. Ernest Pontifex's mother and father, Theobald and Christina, certainly fail to match up to any Victorian ideal of doting parenthood:

> Theobald had never liked children. He had always got away from them as soon as he could, and so had they from him; oh, why, he was inclined to ask himself, could not children be born into the world grown up? If Christina could have given birth to a few full-grown clergymen in priest's orders – of moderate views, but inclining rather to Evangelicalism, with comfortable livings and in all respects facsimiles of Theobald himself – why, there might have been more sense in it; or if people could buy ready-made children at a shop of whatever age and sex they liked, instead of always having to make them at home, and to begin at the beginning with them – that might do better, but as it was he did not like it.

Philip Larkin was even harsher in his judgement of the parent–offspring relationship:

> They fuck you up, your mum and dad.
> They may not mean to, but they do.
> They fill you with the faults they had
> And add some extra, just for you.

And Then the Lover

Adolescence marks another important transition as the child suddenly becomes taller and stronger. The growth of breasts or male genitals may be a source of pride. But muscles and, more worryingly, fat develop in new places; hair grows where it has not grown before; and once-smooth skin erupts in unwanted spots. Both boys and girls in affluent societies begin to spend time and money trying to improve their faces, figures and clothed appearance.

Despite the awkwardness that many feel, normal adolescence

is not necessarily marked by frequent quarrels with parents.[25] Some manage the transition smoothly. Adolescence may, however, be marked by turbulent emotions and uncertainty about identity. John Keats, in his preface to *Endymion*, put it like this:

> The imagination of a boy is healthy, and the mature imagination of a man is healthy; but there is a space of life between, in which the soul is in a ferment, the character undecided, the way of life uncertain, the ambition thick-sighted: thence proceeds mawkishness.

After sexual maturity, humans devote considerable time and energy to courtship and sexual behaviour with the other sex (or, in some cases, with the same sex). Young men roaming around in groups are common enough. In many other species, young males are unable to breed because they are not yet strong enough or experienced enough. In wild horses, for instance, the young adult males form bachelor herds.

Young human males, and increasingly females too, often live their lives as if they have no future. Parents who have given up smoking and other forms of self-destructive behaviour are maddened to find that their children expose themselves to the same risks, apparently in the full knowledge that their lives will be endangered. Whether the risks are taken with smoking, drug abuse, on motorbikes or in unprotected sex, a rational consideration of the consequences seems to bother the young human little at this stage. As Shakespeare wrote in *The Winter's Tale*:

> I would there were no age between sixteen and three-and-twenty, or that youth would sleep out the rest; for there is nothing in the between but getting wenches with child, wronging the ancientry, stealing, fighting . . .

Homicide, particularly in males, peaks sharply at this stage of the lifespan.[26] Such risk-taking may be analogous in some ways to play behaviour – a way of testing and extending the boundaries of the adolescent's world. It may also be a display of their capacity to cope with challenges – a display which is part of competing for a desirable member of the opposite sex. Here again is Shakespeare on the seven ages:

> And then the lover,
> Sighing like furnace, with a woeful ballad
> Made to his mistress' eyebrow. Then a soldier,
> Full of strange oaths, and bearded like the pard,
> Jealous in honour, sudden and quick in quarrel,
> Seeking the bubble reputation
> Even in the cannon's mouth.

At some point, most people realise that they have grown up. Adulthood for many people marks the start of true independence, while others mourn the passing of the childish spirit. 'Maturity', wrote Tom Stoppard, 'is a high price to pay for growing up.' For many people, the search for personal identity continues long into adulthood. W.H. Auden wrote this about the adult years:

> Between the ages of twenty and forty we are engaged in the process of discovering who we are, which involves learning the difference between accidental limitations which it is our duty to outgrow and the necessary limitations of our nature beyond which we cannot trespass with impunity.

The life of the now not-so-young man or woman will probably start to be concerned with reproduction, as opposed to simply having sex. Suddenly the future starts to matter more. Care is taken over a choice of mate and, in due course, parental behaviour flowers. Eight out of ten people in developed societies have children at some point in their lives. The pattern is changing, however, with a growing proportion not having children. For instance, among all women in England and Wales born in 1964, about 40 per cent were still childless at the age of thirty.[27] For those who do it, caring for children requires much energy and many resources. Budgeting of time takes on a completely different significance because so much has to be done. We shall consider some of the issues connected with choosing a mate in Chapter 12.

And Then the Justice

Time marches on, and middle age approaches. Andrew Marvell, in his love poem 'To His Coy Mistress', expressed the impatience that comes when lust is combined with the mature adult's intimations of mortality:

> But at my back I always hear
> Time's wingèd chariot hurrying near:
> And yonder all before us lie
> Deserts of vast eternity.

About the time that children grow up and have their own offspring, their mothers become infertile and the level of the male hormone testosterone in their fathers' bloodstreams drops sharply.[28] At about the same time, their parents' hair starts to turn grey and drop out, their skin starts to wrinkle, their waists thicken and their muscles start to lose their tone. Less noticeably, hair starts to sprout from their eyebrows, ears and noses. Generally they look different from younger people. Shakespeare is caustic:

> And then the justice,
> In fair round belly with good capon lined,

With eyes severe, and beard of formal cut,
Full of wise saws and modern instances;
And so he plays his part. The sixth age shifts
Into the lean and slippered pantaloon,
With spectacles on nose and pouch on side,
His youthful hose well saved, a world too wide
For his shrunk shank; and his big manly voice,
Turning again towards childish treble, pipes
And whistles in his sound.

At this stage both sexes may provide knowledge and leadership for their own society. The roles which the senior members of a society adopt, both in caring for their own and defending them from others, may be assisted by the unmistakable badges they now wear, such as their grey hair and redistributed fat. The dominant male gorilla develops a silver back which signals his status, and primatologists like to use the same term – silverback – for older examples of the gorilla's cousin, the human.

Women live for up to fifty years after they have ceased to be able to reproduce. In other primate species, in contrast, females invariably die before or soon after menopause. Humans are unique in their long post-menopausal lifespan. In middle-aged women, the menopause results from an orchestrated shutting down of reproductive capacity, and therefore looks as though it has been adapted by Darwinian evolution for a specific function. Women can continue to influence the extent to which their genes are represented in future generations even after they have become infertile. They do so by helping to care for their grandchildren. Menopause may have evolved because, after a certain age, it pays the woman more in biological terms to care for her grandchildren than to attempt once again to breed herself.[29]

This theory is supported by evidence from studies of pre-industrial human tribes such as the Hadza of Tanzania.[30] Human infants depend on care from their mother and other adults. The child depends on others to provide food for years after weaning. In the Hadza and other tribes it is common for post-menopausal women to carry on gathering food, some of which is shared with their grandchildren. Their behaviour enhances their grandchildren's ability to survive and reproduce, and will therefore be favoured by Darwinian evolution.

The age-related changes in sexuality that accompany menopause are not unique to women.[31] Indeed, it is increasingly common for the 'male menopause' to be treated with testosterone in much the same way as oestrogen hormone replacement therapy is used to treat some menopausal women.[32] The drug Viagra has been developed to cope with the impotence that embarrassingly confronts some men who perhaps imagined that their sex lives would remain unchanged. The bodily changes are real, however, and may have a biological function. They may reflect the need to shift priorities in those who have the knowledge and experience to lead and protect their own social groups.

Last Scene of All

Finally to Shakespeare's seventh age (and our eighth):

> Last scene of all,
> That ends this strange eventful history,
> Is second childishness, and mere oblivion,
> Sans teeth, sans eyes, sans taste, sans everything.

Awareness of mortality and the old age that precedes it has always occupied human thought. Virgil was gloomy in his attitude towards old age and death:

> All the best days of life slip away from us poor mortals first;
> illnesses and dreary old age and pain sneak up,

and the fierceness of harsh death snatches away.

In humans in industrialised countries the death rate at the age of sixty is ten times the death rate at the age of thirty. An average sixteen-year-old has a one in 2,000 chance of dying within the next year, whereas a hundred-year-old has a one in two chance of dying. For those who survive long enough, senescence brings shrinkage of bones, higher blood pressure and a host of minor and major ailments that eventually lead to death. The age at which these markers of senescence appear is affected by diet, and they show up earlier in smokers, heavy drinkers and those with stressful jobs.[33] With increased attention given to health care and lifestyle in developed countries, longevity continues to rise. For this reason, perhaps, medicine has traditionally dodged the question of the evolutionary significance of senescence, and has looked for a cure as though immortality were an option. Has biology anything to add to a topic that leaves most people frightened or bemused?

The Darwinian analysis of old age is couched in terms of life-history strategies.[34] After the initial high mortality of early life, the probability of surviving from one year to the next is roughly constant for many adult birds and mammals living in natural habitats. This is because random accidents and disease affect individuals of all ages to more or less the same extent. The likelihood of surviving from one year to the next is the same, irrespective of age. The result is that few if any individuals are found alive beyond a certain age and no upwards surge in mortality occurs in older animals. Senescence is barely visible under natural conditions.

So it may have been for humans. Before the advent of modern civilisations, everyone was dead before they were senile. As a result, genes that contribute to the modern diseases of old age, such as cancer and heart disease, were not removed by Darwinian evolution. They are only expressed after many years of life. Evolutionary pressure could not operate against genes whose actions only became manifest after the individual had already reproduced and cared for grandchildren, let alone after most people had died. Similarly, genes that exerted beneficial influences early in life, but which produced bodily malfunctioning in later life, remained in the gene pool. It did not matter if cellular control and repair mechanisms would have stopped working, causing cancer, because the individual was probably dead anyway. Conversely, nobody lived long enough to benefit from genes that would have contributed to a longer life and further opportunity to reproduce or care for grandchildren. An upper limit may nevertheless have been set by biology because the benefits of constant repair had to be offset by the costs.[35]

For many the human world is now a totally different place, of course. Ageing, as it is now experienced in industrialised countries, was not part of the natural life cycle. During the twentieth century, life expectancy in developed countries has almost doubled. The proportion of elderly people in the population is growing rapidly and, if such trends continue, the majority of the population in the developed world will be more than fifty years old by around 2030.

Evolutionary thinking about senescence strongly suggests that growing old is a process of piecemeal breakdown in different parts of the body. Humans die of old age because bodies gradually accumulate flaws, rather than because they have been programmed to stop working at a certain age. However, some evidence points to a degree of unitary control over certain aspects of the ageing process. People suffering from a rare inherited disorder known as Werner's syndrome age prematurely.[36] By their twenties, sufferers have become grey-haired, wrinkled, and prone to diseases of old age such as cancer, osteoporosis, cataracts and coronary heart disease. The symptoms of premature ageing start to appear at around puberty and death generally occurs in the mid-forties, usually from heart attack or cancer.

Werner's syndrome sufferers are believed to lack a gene that is required to make the enzyme helicase.[37] This enzyme helps to correct errors in DNA, a process of crucial importance since new proteins required to maintain the body are constructed throughout life. Without helicase, genetic errors accumulate more rapidly than normal, resulting in premature ageing. If a single faulty gene underlying Werner's syndrome produces a premature plunge into senescence, could the same gene become faulty in normal people in late middle age? If so, the prospect would be opened up of helping the gene to remain functional for longer. Such speculation aside, it is highly unlikely that an immortality drug will ever be found that will remove the spectre of the decaying senses, skeleton, musculature and mind that everyone must expect to accompany their old age.

Those who seek to prolong their lives for as long as possible, even in the face of senescence, should consider the plight of the Struldbruggs in Jonathan Swift's *Gulliver's Travels.* In the Kingdom of Luggnagg a child is occasionally born with a red spot on its forehead – an infallible sign that the child is an immortal, or Struldbrugg, and will never die. When Gulliver first discovers the Struldbruggs he muses on the presumed delights of immortality. They must, he assumes, be happy beyond all comparison, these fortunate beings 'who being born exempt from that universal Calamity of human Nature, have their Minds free and disingaged, without the Weight and Depression of Spirits caused by the continual Apprehension of Death'. But Gulliver is wrong, as he soon finds out from his Luggnuggian hosts. Gulliver sees the true predicament of the immortal Struldbruggs:

> they commonly acted like Mortals, till about Thirty Years old, after which by Degrees they grew melancholy and dejected,

increasing in both till they came to Fourscore . . . When they came to Fourscore Years, which is reckoned the Extremity of living in this Country, they had not only all the Follies and Infirmities of other old Men, but many more which arose from the dreadful Prospect of never dying. They were not only opinionative, peevish, covetous, morose, vain, talkative; but uncapable of Friendship, and dead to all natural Affection, which never descended below their Grand-children . . .

At Ninety they lose their Teeth and Hair; they have at that Age no Distinction of Taste, but eat and drink whatever they can get, without Relish or Appetite. The Diseases they were subject to, still continue without encreasing or diminishing. In talking they forget the common Appellation of Things, and the Names of Persons, even of those who are their nearest Friends and Relations . . .

They were the most mortifying Sight I ever beheld; and the Women more horrible than the Men. Besides the usual Deformities in extreme old Age, they acquired an additional Ghastliness in Proportion to their Number of Years, which is not to be described; and among half a Dozen I soon distinguished which was the eldest, although there were not above a Century or two between them.

The Reader will easily believe, that from what I had heard and seen, my keen Appetite for Perpetuity of Life was much abated. I grew heartily ashamed of the pleasing Visions I had formed; and thought no Tyrant could invent a Death into which I would not run with Pleasure from such a Life.

A positive, mould-breaking approach to old age and death has helped many people to live happy and fulfilled lives when they might otherwise have settled into hopeless inactivity. Awareness that they are able to change long-standing habits and derive pleasure from new experiences makes all the difference. It does not imply a lapse into the fantasy of supposing that it is possible to live for ever. One bright-eyed old lady put it well, 'I hope that when I die I shall have a full diary.'

3

The Sparks of Nature

How hard it is to hide the sparks of nature!
William Shakespeare, *Cymbeline*

The Family Face

A psychologist once wrote that his colleagues tended to emphasise the overwhelming importance of the environment in human development – that is until they had their second child. Learning – which is one of the more obvious ways in which an individual's behaviour is affected by environment and experience – has been a major preoccupation of twentieth-century psychology. And yet, as every parent soon discovers, when all else seems to be equal, siblings are different from each other. Not that all else *is* equal, of course: parental skills expand with experience, anxieties about pregnancy diminish, and the presence of an older sibling makes a big difference to the world in which the second-born develops. These are important points to which we shall return. Nonetheless, the surprisingly large differences between siblings perplex and delight their parents. Since the nurturing environment appears to be broadly the same, some parents see the differences as stemming from the unique genetic nature of each child. The conversion experience may mean that, having believed wholeheartedly in the overwhelming importance of nurture, the parents become convinced instead by the pervasive power of genetic influences.

How do scientists analyse the ways in which genetic and environmental influences leave their mark on each individual during development? Advocacy is rarely dispassionate, particularly when it comes to the nature–nurture debate. It is therefore important to take a cool look at how differences in behaviour, personality, physical growth and other individual distinguishing characteristics may be attributed to any causal factor, whether it be genes or experience.

Thomas Hardy wrote of heredity:

> I am the family face;
> Flesh perishes, I live on,
> Projecting trait and trace
> Through time to times anon,
> And leaping from place to place
> Over oblivion.

A child's characteristics are not a simple blend of its parents' characteristics. Most parents will find some particular likeness between themselves and their child. A daughter might have her mother's hair and her father's shyness, for instance. The child may also have characteristics found in neither parent: a son may have the jaw of his grandmother and the moodiness of his cousin. The shuffling of discrete and supposedly inherited characteristics from one generation to the next is a commonplace of conversation. Not thinking of such shuffling, a woman is said to have written to George Bernard Shaw, 'You have the greatest brain in the world and I have the most beautiful body; so we ought to produce the most perfect child', to which Shaw is said to have replied, 'Yes, but fancy if it were born with my beauty and your brains?'

Spirally Bound

Understanding what happens as inherited characteristics are shuffled during reproduction was one of the great scientific advances. It started with the initially obscure research of the Austrian monk Gregor Mendel in the mid-nineteenth century. Mendel knew nothing of the molecular details, but he was able to formulate rules of inheritance which, when his work was later rediscovered, led to the establishment of the new subject of genetics. Deductions about the particles of inheritance, the genes, continued apace without anybody knowing what they looked like. By degrees, the genes were found to be situated on the chromosomes inside the nuclei of all cells of the body (except mammalian red blood cells, which do not have nuclei) and the molecule of inheritance was later identified as deoxyribonucleic acid (DNA). Eventually, nearly a hundred years after Mendel's scientific papers were published, the double-helical structure of DNA was deduced by James Watson and Francis Crick in Cambridge. As the Australian poet Les Murray put it: 'life's slim volume spirally bound'.

The agents of genetic inheritance, the genes, come in pairs. They separate when sperm or eggs are formed and then recombine at fertilisation. The genes are arranged along 23 pairs of chromosomes in humans. With one exception, the paired chromosomes look alike. The exception is that, in males, one of the chromosomes (the Y) is smaller than its partner (the X) and has at least 115 fewer genes on it. Females have two similar X chromosomes in place of the male XY complement.

In the 22 chromosome pairs that do resemble each other in males, and the 23 chromosome pairs in females, each gene may or may not be the same as its opposite number on the partner chromosome. When the two genes differ from each other, one may be dominant over the other (which, in contrast, is known as recessive). This means that only one of the genes in the pair, the dominant one, affects a characteristic of the organism, such as its eye colour or the structure of a protein. In a few cases the influence of one gene on another may be determined not by the dominance of the gene itself but by the sex of the parent from which it comes – a phenomenon known as genomic imprinting.[1] Usually, the recessive gene remains unexpressed unless it is paired with another gene that is also recessive, in which case it does exert an effect on development. Since the genetic contribution from one of the parents may be different, inherited characteristics may skip a generation. That is why children sometimes resemble their grandparents more closely than they do their parents.

Plant and animal breeders know well that many of the characteristics which matter to them are inherited, in the sense that a new set of progeny will resemble individuals in the ancestral pedigree of that plant or animal more than they resemble progeny from some other pedigree. Long before genes and DNA were discovered, breeders took this as a bountiful fact of life, even though they had no idea how inheritance worked.

For centuries, and in some cases millennia, domestic animals have been artificially selected by humans for breeding because they exhibit specific physical or behavioural features which are regarded as desirable. Dogs, in particular, have for many centuries been bred for their behavioural characteristics. The sheepdog is especially sensitive to the commands of humans, waiting until the shepherd gives it a signal to start herding the sheep. Another breed, the pointer, behaves in a way that helps in sports shooting. When the pointer detects the smell of a game species such as a grouse, the dog stops in its tracks, stiffly orientated towards the bird. Valued behavioural characteristics such as these are clearly inherited, do not need to be taught (at least, not in their most basic form) and are quickly lost if breeds are crossed with others. Humans may also reveal through their children how particular characteristics are inherited. Two healthy parents from a part of the world where malaria is rife may have a child who develops severe anaemia. Both parents carry a gene that does have some effect on red blood cells, protecting them against the malarial parasite which enters red blood cells for part of its life cycle.[2] However, a double dose of this recessive gene leads to the red blood cells collapsing from their normal biconcave disc shape into strange sickle-like shapes. The child who receives this genetic legacy has sickle-cell anaemia.

Few behavioural characteristics are inherited in as simple a fashion as sickle-cell anaemia, and, when they are, the effects are usually damaging and pervasive. A well-known case is the disabling disease phenylketonuria (PKU). If a child inherits two copies of a particular recessive gene from both its parents, it cannot produce a crucial enzyme required to break down phenylalanine, an amino acid which is a normal component of the average diet. The resulting accumulation of phenylalanine in the body poisons the child's developing brain and causes severe mental retardation – unless the condition is diagnosed and the child is given a special diet. Phenylketonuria has clear-cut and relatively simple genetic origins, and the biochemical mechanism by which the responsible genes exert their damaging effects is well understood – in marked contrast to most other instances of inborn errors of metabolism. Nevertheless, the best remedy at present for this genetic defect is to ensure that the child's diet contains no phenylalanine – in other words, to change the child's environment.

A change in a single gene may sometimes have a specific effect on behaviour. In the fruit fly *Drosophila*, one of two paired genes is associated with a distinct form of foraging known as 'sitter' behaviour.[3] An alternative form of the same gene is associated with a distinct alternative style of foraging known as 'rover' behaviour. Larvae which inherit two copies of the recessive *sitter* gene tend to feed in the same spot, while those with one or more *rover* genes move around more, even when food is nearby. Roving behaviour consumes energy, however, and must therefore carry a compensatory advantage. Experiments have shown that *rover* individuals are at an advantage when the population density is high and competition for food is consequently intense. Conversely, the *sitter* behaviour pattern is beneficial – and becomes more common within successive generations – at low population densities. The larvae crawl on average twice as far when the population has been at a high density for many generations, showing that roving behaviour has been more beneficial and hence the *rover* gene has become more frequent in the population.

Sometimes, even in humans, particular behavioural characteristics have been linked with characteristics of a particular chromosome or set of genes. The disorder known as Turner's syndrome affects only girls, and arises when all or part of one X sex chromosome is missing. Turner's syndrome sufferers are typically short, sexually underdeveloped and, although they are usually of normal intelligence, their social behaviour is often limited. In particular, they lack flexibility and responsiveness in social situations and have a lower verbal intelligence than normal girls.[4] In this respect they are more like boys.

Observations such as these, where known features of an individual's genetic make-up can be correlated with aspects of their behaviour, help to narrow down the search for specific genetic influences on behaviour. Of course, correlations can be misleading, because an unknown third factor may affect both of the things that have been measured. Even so, as molecular biology and genetics advance, more and more cases are being found where one or more of the crucial genes is located, identified, and then sequenced so that its precise molecular structure becomes known.

Evidence for genetic influences on human behaviour is usually indirect. It is bound to be so because naturally occurring breeding experiments are rare, and deliberate breeding experiments in the interests of genetic research would obviously be intolerable in most societies. What is more, many genes are involved in the great majority of family likenesses, whether physical or behavioural. However, some light is cast on the links between genes and behaviour by the study of twins.

Twins

The extraordinary similarities between twins, and especially the almost mystical bond between identical twins, is a well-worked theme in imaginative literature. Bruce Chatwin's novel *On the Black Hill*, for example, is the tale of Lewis and Benjamin Jones, identical twins inhabiting a Welsh hill farm. For forty-two years they sleep side by side in their late parents' bed. In their middle age they encounter a woman who has studied twins and is particularly fascinated by Lewis and Benjamin:

> Twins, she said, play a role in most mythologies. The Greek pair, Castor and Pollux, were the sons of Zeus and a swan, and had both popped out of the same egg:
>
> 'Like you two!'
>
> 'Fancy!' They sat up.
>
> She went on to explain the difference between one-egg and two-egg twins; why some are identical and others not. It was a very windy night and gusts of smoke blew back down the chimney. They clutched their heads as they tried to make sense of her dizzying display of polysyllables, but her words seemed to drift towards the borderland of nonsense: '. . . psychoanalysis . . . questionnaires . . . problems of heredity and environment . . .' What did it all mean? At one point, Benjamin got up and asked her to write the word 'monozygotic' on a scrap of paper. This he folded and slipped in his waistcoat pocket.
>
> She wound up by saying that many identical twins were inseparable – even in death.
>
> 'Ah!' sighed Benjamin in a dreamy voice. 'That's as I always felt.'

Research into the inheritance of human behaviour has been greatly helped by comparing genetically identical twins with non-identical twins. Identical (or monozygotic) twins are genetically identical because they are derived from the splitting of a single fertilised egg. They are naturally occurring clones. Non-identical (or dizygotic) twins, in contrast, develop from two fertilised eggs. Consequently, they are no more similar to each other genetically than any two siblings born at different times. In the Western world, about one in eighty-three births are twins, of which a third are identical twins. Giving birth to non-identical twins is particularly likely in certain families. In an isolated community in south-west Finland on the Åland and Åboland archipelago, 21 per cent of women gave birth to twins.[5] In general twinning is most common in peoples of African origin and rare in people of Asiatic origin. No such patterns of inheritance are found in the incidence of identical twins.

If identical twins are no more alike than non-identical twins in a given behavioural characteristic, then this suggests that the genetic influence on that characteristic is weak. Conversely, when identical twins are substantially more alike than non-identical twins (or siblings) then the mechanism of inheritance is likely to be through the nuclear genes. We must add, though, that other mechanisms of inheritance are known, such as genes found in the mitochondria, the biochemical powerhouses of each cell. We shall return to a discussion of inheritance later, since it is wrong to suppose that inheritance simply means genes.[6]

Another way of exploring how genes influence behaviour is to compare twins who have been reared apart (because one or both of them has been adopted soon after birth) with twins who have been reared together. The thought behind this approach is that separation in early infancy removes the influence of the shared environment, leaving only the inherited factors. The thought is not wholly correct, however, because even twins who are separated immediately after birth will have shared a common environment for the first crucial nine months after conception, while they are together in their mother's womb.[7] This obvious truth can add to the difficulties of sorting out the sources of individual distinctiveness. Moreover, being separated at birth and raised in environments that are assumed to be different does not preclude the possibility that their environments may in fact have many important features in common.

Nevertheless, the appearance, behaviour and personality of identical twins who have been reared apart are often startlingly similar. In one documented case, for example, a pair of twins had been separated early in life, one growing up in California, the other in Germany. Yet when they met for the first time in thirty-five years, they both arrived wearing virtually identical clothes and with similar clipped moustaches; both had a habit of wrapping elastic bands around their wrists; and both had the idiosyncratic habit of flushing lavatories before as well as after using them.[8]

Accounts such as these are sometimes greeted with scepticism because it is suspected that only the startling matches have been reported while the discrepant twins have been ignored in the interests of a good story. Nevertheless, some properly conducted statistical surveys have revealed that, on a range of measures of personality, identical twins who have been reared apart are more like each other than non-identical twins also reared apart.[9] When making such comparisons it does not matter whether, as has often been argued, the measures of behavioural characteristics are crude and relatively insensitive. While differences are less likely to be found with insensitive behavioural measures, differences *are* found. The inescapable conclusion is that some observable aspects of individuals' behaviour are influenced by inherited factors.

Further evidence for genetic influences comes from observing patterns of behavioural development. Many aspects of behaviour and intellectual functioning do not develop at a steady rate:

obvious mental 'growth spurts' are found in children's capacities for complex thought and reasoning, and individual children differ in the chronological ages at which these developmental spurts occur.[10] As with many other behavioural characteristics, identical twins are more alike in the timing of such developmental spurts than are non-identical twins. The implication is that individual differences in the patterns of behavioural development themselves are influenced by inherited factors.

What About the Environment?

Genes matter. But even the most cursory glance at humanity reveals the enormous importance of each person's experience, upbringing and culture. Look at the astonishing variation among humans in language, dietary habits, marriage customs, child-care practices, clothing, religion, architecture, art and much else besides. Nobody could seriously doubt the remarkable human capacity for learning from personal experience and learning from others. Or so you might think. Yet in 1994 Richard Herrnstein and Charles Murray wrote in their notorious but widely read book *The Bell Curve*: 'Success and failure in the American economy are increasingly a matter of the genes that people inherit . . . Programmes to expand opportunities for the disadvantaged are not going to make much difference.' They were wrong. Early intervention can benefit the disadvantaged child, but in ways that had not been fully anticipated. In the 1960s, great efforts were made in the United States to help people living in difficult and impoverished conditions. A large government programme known as Headstart was designed to boost children's intelligence by giving them educational experience before starting school. In the event, the Headstart programme did not seem to have the substantial and much-hoped-for effects on intelligence, as measured by IQ.[11] Children who had received the Headstart experience displayed an initial, modest boost in their IQ scores, but these differences soon evaporated after a few years. The fashionable response was to disparage such well-meaning efforts to help the disadvantaged young.

Later research, however, has revealed that some of the other effects of the Headstart experience were long-lasting and of great social significance – greater, in fact, than boosting IQ scores. Several long-term follow-up studies of children who had received pre-school training under Headstart found that they were distinctive in a variety of ways, perhaps the most important being that they were much more community-minded and less likely to enter a life of crime.[12] Headstart produced lasting benefits for the recipients and society more generally, but not by raising raw IQ scores. Evidence for the long-term benefits of early educational intervention has continued to accumulate.[13] Studies like these raise many questions about the ways in which early experiences exert their effects, but they do at least show how important such experiences can be.

Even relatively subtle differences in the way children are treated at an early age can have lasting effects on the way they behave years later. One study compared the long-term effects of three different types of pre-school teaching.[14] In the first type, three- and four-year-olds were given direct instruction, with the teacher initiating the children's activities in a strict order. The second type of teaching was a traditional nursery school in which the teachers responded to activities initiated by the children. In the third, known as High/Scope, the teachers involved the children in planning their own activities, but arranged the classroom and the daily routine so that the children could do things that were appropriate to their stage of development.

Striking differences were found between the children as they grew up. When followed up at the age of twenty-three, the individuals who had been in the direct instruction group were worse off in a variety of ways than those in the other two groups. In particular, they were more likely to have been arrested on a criminal charge and more likely to have received special help for emotional impairment. In comparison, people who had received the more relaxed type of pre-schooling were more likely to be living with spouses and much more likely to have developed a community spirit.

Early Handling

Experience does not exert its effects only through learning, however. Big, long-lasting differences in the adult behaviour of individuals can be generated by other environmental influences that do not appear to involve learning at all. An example from animal studies – the phenomenon of early handling – shows how important these influences can be.

Merely handling young rat pups can modify their physiological and behavioural response to stressful conditions encountered much later in their adult life.[15] Laboratory-reared pups that have been exposed to brief daily periods of handling by a human during the first few weeks of life subsequently exhibit a quite different hormonal response to stress, in comparison with rats that have not experienced early handling. When they are exposed to a stressful stimulus in adulthood, such as being physically restrained, early-handled rats show a sharp and rapid release of adrenocorticotrophic hormone (ACTH) from the pituitary gland, situated at the base of the brain. This hormone in turn triggers a pulse of corticosteroid hormones from the adrenal glands. Corticosteroids, the archetypal 'stress hormones', prepare the animal's body to deal with stressful challenges in the environment.

Non-handled animals, by contrast, show a much slower increase in ACTH, and their corticosteroid level, once raised in response to stress, stays high for much longer.[16] Their physiological response is much less well tuned for dealing with stressful challenges. Non-handled rats are less able to cope with novelty or stress, and remain in a fearful state for much longer than handled rats. Prolonged elevations in corticosteroid levels can have damaging effects on the body, including suppressing the immune system and damaging the nervous system, where persistently high levels of cortisol accelerate the loss of certain types of neuron during the ageing process. Consistent with these effects of corticosteroids, early-handled rats show slower neural degeneration of the brain and are better able to learn new tasks in old age. Physiological differences between handled and non-handled rats are still present after two years, which is a good age for a rat, suggesting that the effects of early handling are highly persistent, if not permanent.[17] What does early handling do to the young animal?

When a rat pup is handled by a human, it gives out ultrasonic distress calls. These distress calls stimulate the pup's mother to lick it more when it is subsequently returned to its home cage. Pups which have been handled receive twice the amount of grooming from their mothers than non-handled pups.[18] The maternal behaviour elicited by the distress calls in turn affects the development of hormonal systems which mediate the pups' subsequent response to stress. Pups whose mothers spontaneously lick them most during their first ten days of life also show smaller hormonal stress responses when adult, as though they have experienced early handling. The more a mother grooms her young pup, the more finely tuned her offspring's responses to stress in adulthood. Maternal care, in the form of licking and grooming, has an important influence on the developing animal's ability to cope with stress when it has grown up.

Normal development of the stress response seems to depend on a form of experience – being licked by the mother – that is usually predictable and stable from one generation to the next under natural conditions. The importance of this experience is only revealed when rats are reared in the artificial conditions of the laboratory. In an unstimulating environment of constant temperature and humidity, with freely available food and water and no predators, some rat mothers do not care for their pups as vigorously as they would under natural conditions. Their pups are, in effect, mildly deprived of maternal care. If, however the pups are handled, their ultrasonic distress calls stimulate the lazier mothers to behave more as they would have done in the natural environment, thereby avoiding the long-term effects of maternal deprivation.

The early-handling story has a delicious twist to it. For many years the scientists who studied the phenomenon were unaware of the ultrasonic calls emitted by the handled pups, and consequently did not appreciate the crucial role of the mother in the whole process. Instead, they regarded the non-handled rats as the 'control group', representing the normal baseline against which the handled rats were compared. In reality, however, the experience of the handled rats was closer to normality. The developmental mechanisms that might have been required to cope with the ill-effects of having a lazy mother presumably never evolved because lazy rat mothers did not exist – at least, until scientists constructed unnatural, super-affluent and very boring laboratory environments for them.

Tempting though it may be to draw simple parallels between early-handled rats and children reared by assiduous or neglectful parents, it is far from clear how relevant the early-handling story actually is to human development. At one time, the effect of early handling on rats was taken as an explanation, if not a justification, for the deliberate scarring of children or the mutilation of their genitals that is practised in some cultures. The link was an idea that steeling with physical punishment makes children better able to cope with stress later in life. While we doubt this is true, one general lesson can be extracted from the early-handling story. It is that many features of the childhood environment, which are usually constant from one family to the next, play a central role in normal development. Their importance only becomes apparent when they are absent.

Dear Octopus

Although peers become ever more important as a child develops, a central aspect of most children's lives is their family – that nexus of people and relationships which Dodie Smith described as 'that dear octopus from whose tentacles we never quite escape'. Like early handling, the family exerts pervasive, long-term effects on the developing child, with consequences that are as profound as they are difficult to predict. Family members form a crucial part of any individual's environment. For example, behavioural differences may arise because, when siblings behave in one way, it becomes advantageous to behave in a different way. An important aspect of the family environment is the child's relative position within it. In a large family, the child who is born first clearly experiences a different world from his or her youngest sibling. Each child therefore experiences what might appear to be the same family environment in quite different ways. As Jane Austen put it, in *Emma*, 'Nobody, who has not been in the interior of a family, can say what the difficulties of any individual of that family may be.'

Evidence has accrued over many years that birth order plays a subtle but influential role in the development of personality, styles of thinking and even athletic prowess. Compared with first-borns, later-borns are more likely to engage in dangerous sports such as rugby, boxing and parachute jumping.[19] This may be due in part to their better health: first-born children are more likely to have allergies,[20] perhaps because infections brought in by the older sibling protect the second-born against allergic disease. First-borns, on the other hand, have marginally higher IQ scores on average than second-borns.[21] Significantly more classical music composers are first-born or only children than would be expected by chance.[22]

A massive twenty-five-year study of family dynamics by the American psychologist Frank Sulloway revealed some remarkable connections between birth order, personality and intellectual style.[23] Sulloway's analysis showed that first-borns are less likely to be unorthodox in their thinking than second- or later-born children. Among intellectuals, those who challenged established doctrines were most commonly second- or later-borns. Out of 800 eminent scientists, 60 per cent of later-borns were found to have supported revolutionary theories, compared with 40 per cent of first-borns. Copernicus, Freud and Darwin, among many other scientific revolutionaries, were later-borns. Birth order also appears to influence the development of sexuality – a subject to which we shall be returning later. Fewer homosexual men are first-borns than would be expected by chance, and they have a significantly greater number of older brothers on average than do heterosexual men.[24]

The associations between birth order and adult characteristics are almost certainly heterogeneous, initiated in different ways, and involve many different mechanisms in development. We shall return to them in Chapter 4. Their interest here, though, is that they occur at all. Like early handling in rats, they reveal some of the numerous and subtle environmental influences which operate on children as they grow up.

Myopia and Music

The importance of both genes and environment to the development of all animals, including humans, is obvious. This is true even for apparently simple physical characteristics, let alone complex psychological variables. Take myopia (or short-sightedness) for example. Myopia runs in families, suggesting that it is inherited. But it is also affected by the individual's experience. Both a parental history of myopia and, to a lesser extent, the experience of spending prolonged periods studying close-up objects will predispose a child to become short-sighted.[25]

A more interesting case is musical ability, about which strong and contradictory views are held. Popular beliefs about the origin of special talents are generally that they are inherited. Thomas Mann suggested in *Confessions of Felix Krull* that 'Natural gifts and innate superiorities customarily move their possessors to a lively and respectful interest in their heredity.' Before getting too deeply into this, it is first necessary to ask whether musical ability is a specific talent at all, or merely the reflection of more general mental and physical capacities. Dissociation between general intellectual capability and musical ability is strongly suggested by the phenomenon of the musical *idiot-savant* – an individual with low intelligence but a single, outstanding talent for music.[26] Such people are usually male and often autistic. Their unusual gift – whether it be for music, drawing or mental arithmetic – becomes apparent at an early age and is seldom improved by practice. One typical individual could recall and perform pieces of music with outstanding ability and possessed almost perfect pitch; he had poor verbal reasoning, but his low intellectual ability was to some degree offset by high levels of concentration and memory. However, children who are good at music also tend to be good at reading and have a good sense of spatial relations, even after taking account of variables such as age and IQ.[27]

The main factors fostering the development of musical ability form a predictable cast: a family background of music; practice (the more the better); practical and emotional support from parents and other adults; and a good relationship with the first music teachers.[28] Practice is especially important, and attainment is strongly correlated with effort. A rewarding encounter with an inspirational teacher may lock the child into years of effort, while conversely an unpleasant early experience may cause the child to reject music, perhaps for ever. Here, as elsewhere, chance plays a role in shaping the individual's development.

Research on identical and non-identical twins has shown that the shared family environment has a substantial influence on the development of musical ability, whereas inherited factors exert only a modest effect. Genetically identical twins are only slightly more alike in their musical ability than non-identical twins or siblings.[29] A study of more than 600 trainee and professional musicians analysed the origins of perfect (or absolute) pitch – that is, the ability to hear a tone and immediately identify the musical note without reference to any external comparison.[30] Heritable factors appeared to play a role, as musicians with perfect pitch were four times more likely than other musicians to report having a relative with perfect pitch. But the same study also found that virtually all the musicians with perfect pitch had started learning music by the age of six. Of those who had started musical training before the age of four, 40 per cent had developed perfect pitch, whereas only 3 per cent of those who had started training after the age of nine possessed the ability. Early experience is also important.

Like many other complex skills, musical ability develops over a prolonged period; and the developmental process does not suddenly stop at the end of childhood. Expert pianists manage to maintain their high levels of musical skill into old age despite the general decline in their other faculties.[31] They achieve this through copious practice throughout their adult life; the greater the amount of practice, the smaller the age-related decline in musical skill. Practice not only makes perfect, it maintains perfect.

How Much Nature, How Much Nurture?

Is it possible to calculate the relative contributions of genes and environment to the development of behaviour patterns or psychological characteristics such as musical ability? Given the passion with which clever people have argued over the years that *either* the genes *or* the environment are of crucial importance in development, it is not altogether surprising that the outcome of the nature–nurture dispute has tended to look like an insipid compromise between the two extreme positions. Instead of asking whether behaviour is caused by genes or caused by the environment, the question instead became: 'How much is due to each?' In a more refined form, the question is posed thus, 'How much of the variation between individuals in a given character is due to differences in their genes, and how much is due to differences in their environments?'

The nature–nurture controversy appeared at one time to have been resolved by a neat solution to this question about where behaviour comes from. The suggested solution was provided by a measure called heritability. The meaning of heritability is best illustrated with an uncontroversial characteristic such as height, which is clearly influenced by both the individual's family background (genetic influences) and nutrition (environmental influences). The variation between individuals in height which is attributable to variation in their genes may be expressed as a proportion of the total variation within the population sampled. This index is known as the heritability ratio. The higher the figure, which can vary between 0 and 1·0, the greater the contribution of genetic variation to individual variation in that characteristic. So, if people differed in height solely because they differed in their genes, the heritability of height would be 1·0; if, on the other hand, variation in height arose entirely from individual differences in environmental factors such as nutrition then the heritability would be 0. More than 30 twin studies, involving a total of more than 10,000 pairs of twins, have collectively produced an estimated heritability for IQ of about 0·5 (ranging between 0·3 and 0·7).[32] Twin and adoption studies of personality measures, such as sociability/shyness, emotionality and activity level, have typically produced heritabilities in the range 0·2 to 0·5.[33]

Calculating a single number to describe the relative contribution of genes and environment has obvious attractions. Estimates of heritability are of undoubted value to animal breeders, for example. Given a standard set of environmental conditions, the genetic strain to which a pig belongs will predict its adult body size better than other variables such as the number of piglets in a sow's litter. If the animal in question is a cow and the breeder is interested in maximising its milk yield, then knowing that milk yield is highly heritable in a particular strain of cows under standard rearing conditions is important.

But behind the deceptively plausible ratios lurk some fundamental problems. For a start, the heritability of any given characteristic is not a fixed and absolute quantity – tempted though many scientists have been to believe otherwise. Its value depends on a number of variable factors, such as the particular population of individuals that has been sampled. For instance, if heights are measured only among people from affluent backgrounds, then the total variation in height will be much smaller than if the sample also includes people who are small because they have been undernourished. The heritability of height will consequently be larger in a population of exclusively well-nourished people than it would be among people drawn from a wider range of environments. Conversely, if the heritability of height is based on a population with relatively similar genes – say, native Icelanders – then the figure will be lower than if the population is genetically more heterogeneous; for example, if it includes both Icelanders and African Pygmies. Thus, attempts to measure the relative contributions of genes and environment to a particular characteristic are highly dependent on who is measured and in what conditions.

Another problem with heritability is that it says nothing about the ways in which genes and environment contribute to the biological and psychological cooking processes of development. This point becomes obvious when considering the heritability of a characteristic such as 'walking on two legs'. Humans walk on fewer than two legs only as a result of environmental influences such as war wounds, car accidents, disease or exposure to teratogenic toxins before birth. In other words, all the variation within the human population results from environmental influences, and consequently the heritability of 'walking on two legs' is zero. And yet walking on two legs is clearly a fundamental property of being human, and is one of the more obvious biological differences between humans and other great apes such as chimpanzees or gorillas. It obviously depends heavily on genes, despite having a heritability of zero. A low heritability clearly does not mean that development is unaffected by genes.

If a population of individuals is sampled and the results show that one behaviour pattern has a higher heritability than another, this merely indicates that the two behaviour patterns have developed in different ways. It does not mean that genes play a more important role in the development of behaviour with the higher heritability. Important environmental influences might have been relatively constant at the stage in development when the more heritable behaviour pattern would have been most strongly affected by experience.

Yet another serious weakness with heritability estimates is that they rest on the spurious assumption that genetic and environmental influences are independent of one another and do not interact. The calculation of heritability assumes that the genetic and environmental contributions can simply be added together to obtain the total variation. In many cases this assumption is clearly wrong. For example, in one study of rats the animals' genetic background and their rearing conditions were both varied; rats from two genetically inbred strains were each reared in one of three environments, differing in their richness and complexity.[34] The rats' ability to find their way through a maze was measured later in their lives. Rats from both genetic strains performed equally badly in the maze if they had been reared in a poor environment (a bare cage) and equally well if they had been reared in a rich environment filled with toys and objects. Taken by themselves, these results implied that the environmental factor (rearing conditions) was the only one that mattered. But it was not that simple. In the third type of environment, where the rearing conditions were intermediate in complexity, rats from the two strains differed markedly in their ability to navigate the maze. The genetic differences only manifested themselves behaviourally in this sort of environment. Varying both the genetic background and the environment revealed an interplay between the two influences.

An overall estimate of heritability has no meaning in a case such as this because the effects of the genes and the environment do not simply add together to produce the combined result. The effects of a particular set of genes depend critically on the environment in which they are expressed, while the effects of a particular sort of environment depend on the individual's genes. Even in animal breeding programmes which use heritability estimates to practical advantage, care is still needed. If breeders wish to export a particular genetic strain of cows which yields a lot of milk, they would be wise to check that the strain will continue to give high milk yields under the different environmental conditions of another country. Many cases are known where a strain that performs well on a particular measure in one environment does poorly in another, while a different strain performs better in the second environment than in the first.

Any scientific investigation of the origins of human behavioural differences eventually arrives at a conclusion that most non-scientists would probably have reached after only a few seconds' thought. Genes and the environment both matter. The more subtle question about how much each of them matters defies an easy answer; no simple formula can solve that conundrum. The problem needs to be tackled differently. The approach we shall adopt in Chapter 4 is to look at the processes of biological and psychological development – the cooking of the ingredients.

4

Cooking Behaviour

The problem with raw ingredients is that you have to cook them.
Raymond Blanc, *Blanc Mange* (1994)

The Great Blueprint Fallacy

In Michael Crichton's science fiction novel *Jurassic Park* and the successful Steven Spielberg films made from it, clones of extinct dinosaurs were reconstructed in full. All that was required was fragments of dinosaur DNA recovered from 65-million-year-old fossilised mosquitoes encased in amber, together with some DNA from frogs to fill in the missing bits. The new versions of the dinosaurs looked exactly like the original animals whose blood had been sucked by the amber-ensnared mosquitoes. Moreover, they *behaved* just like the originals too. The genes provided the complete blueprint for the dinosaurs' body, brain and behaviour. The assumption behind *Jurassic Park* is that the whole organism, including its behaviour, is like a Japanese paper flower: simply put the dry paper into water (or, in the case of the story, a suitable growth medium) and it will open out. Among those dinosaur genes, by implication, were genes coding for the animals' complete behaviour patterns. The idea is fantasy.

Sometimes genes are likened to the blueprint for a building.

The idea is hopelessly misleading because the correspondences between plan and product are not to be found. In a blueprint, the mapping works both ways. Starting from a finished house, the room can be found on the blueprint, just as the room's position is determined by the blueprint. This straightforward mapping is not true for genes and behaviour, in either direction. Yet hardly a week goes by without the reporting of a new and supposedly direct link between genes and behaviour. Intelligence, criminality, a desire for novelty, homosexuality – all have their gene. *Gattaca*, another Hollywood movie, was launched with the slogan: 'There is no gene for the human spirit.' Quite right; but the implication was that, while no single gene codes for the human spirit, genes for characters such as being clever or having a pretty face do exist.

The language of a gene 'for' a particular behaviour pattern, so often used by scientists, is exceedingly muddling to the non-scientist (and, if the truth be told, to many scientists as well). What the scientists mean (or should mean) is that a genetic difference between two groups is associated with a difference in behaviour. They know perfectly well that other things are important and that, even in constant environmental conditions, the developmental outcome depends on the whole gene 'team'. Particular combinations of genes have particular effects, in much the same way as a particular collection of ingredients may be used for a particular dish; a gene that fits into one combination may not fit into another. Unfortunately, the language of genes 'for' characters has a way of seducing the scientists themselves into believing their own sound-bites.

When confronted with the question of where behaviour comes from, it sometimes seems as though the world is divided into two groups, the analysts and the chefs. The analysts are like health inspectors who want to know what is *really* responsible when a case of food poisoning is reported. They are preoccupied with tracking down causes and they do their job extremely well.

The chefs, in contrast, want to know how things work. What has to be done to make a wonderful dish? To understand why the *Jurassic Park* and *Gattaca* fantasies are just that – fantasies – requires the chef's understanding of what genes really do.

Genes Make Proteins, Not Behaviour

Each human has about 60,000 different genes, each of which is an inherited molecular strand (or set of strands) which may be translated into a protein molecule (or part of one) in the living organism. Each protein molecule in the individual's body is a specific combination of amino acids. The components of DNA, known as nucleotide bases, are arranged along the strand. Each strand matches another one and the pair form a double helix. A sequence of three nucleotide bases codes uniquely for one of twenty different amino acids that provide the constituents of all proteins found in living organisms. Once formed, the proteins are crucial collectively to the functioning of each cell in the body. Some of the proteins are enzymes, controlling biochemical reactions; others form structures that give an architecture to the whole cell. These protein products of genes do not work in isolation, but in a cellular environment created by the conditions of the local environment and by the expression of other genes. Each gene product interacts with many other gene products.

The adult human body is made up of a hundred million million (10^{14}) cells. Each cell has in its nucleus the full complement of genes inherited from both the mother and the father (except the red blood cells, which do not have a nucleus). About 350 different types of cell are found in the human body and most cells have over 2,000 different types of protein molecule. The same complete set of genes is present in virtually all cells, but they are activated or inactivated at different stages in development. Genes are switched on and off in response to local conditions within the cell. The non-genetic material provided by the mother also plays an important part at the beginning of development. It is possible to remove the nucleus of a fertilised egg of one mouse strain and replace it with the nucleus of a fertilised egg of another mouse strain. When that is done, some genes that would not have been activated are now expressed. These genes in turn affect the characteristics of the whole organism, or what is known as its phenotype.

As the cells of a fertilised egg divide to form the early embryo, the activities of adjoining cells – and, increasingly, the conditions in the environment surrounding the embryo – participate in what looks like a symphony of gene expression. Whether the equivalent of a musical score for this seemingly orderly process will ever be found is another matter. Nevertheless, the discovery of the extent to which one gene triggers other genes has been one of the triumphs of modern developmental biology.[1]

If this were not bad enough for the analysts seeking simple developmental explanations for behaviour, the idealised notion of a gene as a single entity is also mistaken. A gene does not consist of a fixed sequence of coded information like the indentations on a compact disc that code for music. Many lengths of DNA may carry no useful information. The parts that do carry useful information are recognised in the translation process, so that the completed code for a protein molecule is stitched together from separate lengths of DNA. The difficulty of identifying a gene is made worse because, in the course of evolution, a given gene may be copied many times and incorporated into different chromosomes to create a family of identical or very similar molecules.

At least half of all human genes are involved in building the brain and nervous system, although many of these same genes are also involved in building other parts of the body. The adult human brain has around one hundred thousand million (10^{11}) neurons, each with hundreds or thousands of connections to other neurons. A diagram of even a tiny part of the brain's connections would look like an enormously complex version of a map of the London Underground system. The brain is organised into sub-systems, many of which are dedicated to different functions which are run separately but are integrated with each other. Since the behaviour of the whole animal is dependent on the whole brain, it will be obvious why it is not sensible to ascribe a single aspect of behaviour to a single neuron, let alone a single gene. The pathways running from genes to neurons and thence to behaviour are long, full of detours, with many other paths joining and many leading away.

Nothing happens in isolation. The products of genes and the activities of neurons are all embedded in elaborate networks. Each behaviour pattern or psychological characteristic is affected by many different genes, each of which contributes to the variation between individuals (a phenomenon known as polygeny). In an analogous way, many different design features of a motor car contribute to a particular characteristic, such as its maximum speed. Conversely, each gene influences many different behaviour patterns (an effect known as pleiotropy). To use the car analogy again, a particular component such as the system for delivering fuel to the cylinders may affect many different aspects of the car's performance, such as its top speed, acceleration and fuel consumption. The effect of any one gene also depends on the actions of many other genes.

Modern technology allows particular genes to be knocked out of action.[1] Numerous experiments on mice have found that an effect on the whole animal of changing one gene is only observed in particular genetic strains – that is, when a particular combination of other genes is also present. Sometimes, when one or other of two genes is knocked out, no change is observed in either case, but a big change is found if both genes are inactivated. Some genes are only expressed in special environmental conditions. For example, a variant of one particular gene in the fruit fly *Drosophila* causes paralysis – but only when the environmental temperature exceeds 29°C.[2] In development, as in cooking a soufflé, a small difference in temperature can make a big difference to the outcome.

It is clear, then, that because of the immensely complex system in which they are embedded, no simple correspondence is found between individual genes and particular behaviour patterns or psychological characteristics. Genes store information coding for the amino acid sequences of proteins; that is all. They do not code for parts of the nervous system and they certainly do not code for particular behaviour patterns. Any one aspect of behaviour is influenced by many genes, each of which may have a big or a small effect. Conversely, any one of many genes can have a major disruptive effect on a particular aspect of behaviour. A disconnected wire can cause a car to break down, but this does not mean that the wire by itself is responsible for making the car move. All these complications have tempted theorists into arguing that the seemingly simple and orderly characteristics of development (such as they are) are generated by dynamic processes of great complexity.[3]

An illustration of the long and indirect path from genes to behaviour is provided by Kallmann syndrome, a genetic condition that afflicts men only.[4] The main behavioural consequence of this genetic defect is a lack of sexual interest in members of the opposite sex. Kallmann syndrome is caused by damage at a specific genetic locus. Cells that are specialised to produce a chemical messenger called gonadotrophin-releasing hormone (GnRH) are formed initially in the nose region of the foetus. Normally the hormone-producing cells would migrate into the brain. As a result of the genetic defect, however, their surface properties are changed and the cells remain dammed up in the nose. The activated GnRH cells, not being in the right place, do not deliver their hormone to the pituitary gland at the base of the brain. Without this hormonal stimulation, the pituitary gland does not produce the normal levels of two other chemical messengers, luteinizing hormone and follicle stimulating hormone. Without these hormones, the testes do not produce normal levels of the male hormone testosterone, and without normal levels of testosterone, the man shows little sign of normal adult male sexual behaviour. As a result, men suffering from Kallmann syndrome have a reduced libido and are not attracted to either sex. Even in this relatively straightforward example, the pathway from gene to behaviour is long, complicated and indirect. Each step along the causal pathway requires the products of many genes and has ramifying effects, some of which may be apparent and some not.

Niche-Picking

Many birds specialise in eating particular types of seeds, but the size of the bill differs greatly between species. Seeds also differ greatly in size, from some that are only a millimetre across to a hazelnut, which is more than ten times the diameter. When they are young, birds experiment with different sizes of seed until they find ones that are particularly well suited to their bills.[5] The young bird learns to pick the food source to which its body is best suited. Similarly, as children grow up they are given copious advice about picking careers that suit their aptitudes. They also actively choose circumstances they find congenial or ones with which they are best able to cope.[6]

The active participation of each individual in choosing its own physical and social environment can have interesting consequences. It means that people who differ consistently in ways that relate to differences in their genes may also predictably pick certain physical and social environments in which to live. The possibility that they might do this has been given the splendid name of niche-picking.[7] It means that individuals with different characteristics, some of which reflect differences in their genes, end up by their own actions experiencing the world in quite different ways.

The various niches into which individuals settle may be forced upon them by their different physiques, for example. One obvious difference between adult humans is their height – a characteristic that is affected both by nutrition and genes. Does an individual's stature affect his or her personality, as is commonly believed? Some evidence suggests that it does. One study, for example, assessed psychological adjustment in a sample of children and adolescents who had been referred for growth hormone treatment because of their short stature. Roughly half the sample had growth hormone deficiency; the cause of shortness in the others was unknown. The children were of average intelligence and from comfortable social backgrounds. Yet a large number of them were underachieving academically and they displayed a higher than expected rate of behaviour problems.[8] Most small children and adults are psychologically well adjusted, however, and it is certainly not the case that being exceptionally short (or exceptionally tall) is necessarily damaging to the psyche. Fighting back on behalf of the short man, Francis Bacon wrote, 'Wise nature did never put her precious jewels into a garret four stories high: and therefore . . . exceeding tall men had ever very empty heads.'

The hero of William Boyd's novel *The New Confessions* encounters a striking example of the connection between physical size and personality during the First World War in the Bantams, a battalion in which every man is under the British army's minimum height of five foot three inches:

> Five very small men – very small men indeed – sat around a tommy-cooker brewing tea. They looked at us with candid hostility. They wore kilts covered with canvas aprons. Their faces were black with mud, grime and a five-day growth of beard. Two of them stood up. The tops of their heads came up to my chest. Neither of them could have been more than five feet tall . . .

> '*Less fuckin' natter more work youse two English bastards.*'
>
> These words came from platoon sergeant Tanqueray, a Bantam, supervising our work party. The top of his head reached my armpit . . . He hated me and Teague, as did the rest of his men. He was five feet two inches, just under the Army minimum. He was bitter enough as it was, missing out on the chance of a regular battalion by one inch, but having two ex-public schoolboys in his platoon seemed almost to have deranged him . . . I became a symbol of the dark genetic conspiracy that had contrived to render him small.

A long debate has revolved around whether boys and girls differ in their behaviour because they are genetically different from each other or because they are treated differently from a young age. By degrees a consensus has formed that, from early in development, boys are on average more assertive and individualistic, while girls tend to be more expressive and interested in personal relationships.[9] Comparable sex differences in behaviour are found in monkeys and apes.

The initial sex differences in humans often sharply reduce as children develop – at least, under some conditions. Under other conditions, however, the differences are amplified by the normal practices of the society in which they grow up and by their own habit of forming single-sex groups in which to play. The boys exclude the girls from their gangs and likewise the girls tease or ignore boys who seek to play with them. This may also be an example of niche-picking, where the niche will differ greatly from one culture to another.

Experience can also work in the opposite direction – uncovering genetic differences rather than washing them away. In one experiment, identical and non-identical twins who had been reared apart were asked to perform a difficult task of eye-to-hand co-ordination, which involved tracking a moving target on a screen with a pen.[10] Skill in this task was found to correlate well with performance in real-life activities such as driving a car. Some individuals showed an immediate aptitude for the tracking task. If one identical twin was good at it then the other one was also likely to do well, suggesting that inherited factors influenced performance. Everybody's performance improved substantially with practice, showing that experience was also important. As they improved, the performance of each identical twin became even more similar to that of the other. The convergence was greater than in the non-identical twins. In this case, practice seemed to reduce the effects of prior learning, thereby allowing individual differences in inherited abilities to become more apparent.

Environmental and inherited factors often work together to produce much larger overall effects than is the case when either factor is present on its own. One study in Sweden looked at children from families with or without a history of criminality.[11] These children had been adopted early in life into families, some of which also had a history of criminality. Coming from a biological family with a criminal history quadrupled the individual's chances of exhibiting criminality, while being adopted from a non-criminal biological family into a criminal family doubled the risk. But the combination of criminality in both the biological and adopting families increased the risk of turning into a criminal by a factor of fourteen – far more than just the product of the biological and familial influences. The combination of inherited and environmental influences was much greater than the effect of each influence operating on its own.

The conventional analytical method, which partitions behavioural variation into genetic and environmental components, may be misleading in a different way. The two major contributors to variation may not simply add or even multiply together to produce their overall effect. For example, the performances of adopted children in tests of cognitive ability are related to those of their adopting parents and their biological parents. Commonly in such studies, both types of parents have independent effects on the children.[12] The effects of the genes (provided by the biological parents) and the effects of the environment (provided by the adopting parents) seem to add together. In the case of IQ scores, for example, each factor accounts for about 10 per cent of the variation in the children's scores.

A common-sense view of what happens is that initially the cognitive abilities of the child are most strongly affected by its biological parents, but that later in development they are increasingly affected by the experiences the child has had with its adopting parents. However, the quality of the exchanges between the adopting parents and the child will depend on the match between their characteristics. A potentially able child who is adopted by dull people might be less stimulated and more frustrated than if he or she had been adopted by lively intelligent people. Conversely, adopting parents who are disappointed by the less able child might provide a less supportive environment than those whose expectations are satisfied by the responsiveness of an able child. Here again, the difference between the child and its adopting parents probably matters, but this time in the reverse direction. One study, for example, found that the bigger the absolute difference in IQ between the biological and adopting parents, the more the child was adversely affected.[13] The difference between the parents and child accounted for as much of the variation in the children as the direct influences of the biological and adopting parents. The consequences of the relationships between adopting parents and the children were not revealed by a simplistic analysis which assumes that what went in is directly related to what comes out. The appropriate analysis was not carried out until a plausible question was asked about the nature of the developmental process. Examples like this highlight the continuous process of exchange between individuals and their environment.

A Picture of Differences

'We were . . . such a company of characters and such a picture of differences,' wrote Henry James about his family. One surprising conclusion to emerge from studies of identical twins is that twins reared apart are sometimes *more* like each other than those reared together.[14] To put it another way, rearing two genetically identical individuals in the same environment can make them less similar rather than more similar because one of the twins is dominant to the other, entering the room first and speaking for them both. This fact pleases neither the extreme environmental determinist nor the extreme genetic determinist. The environmental determinist supposes that twins reared apart must have different experiences and should therefore be more dissimilar in their behaviour than twins who grew up together in the same environment. The genetic determinist does not expect to find any behavioural differences between genetically identical twins who have been reared together. If they have had the same genes and the same environment, then how can they be different?

Stories about the deep empathy between twins make cases in which identical twins diverge from each other seem all the more remarkable. Or rather, it means that when they go their different ways they behave like other children. Children seek out their own space. When Mary did well at art, her younger sister Susan would not have anything to do with drawing or painting, even though she would probably have been good at both. When Henry developed a flair for history and languages, George inclined towards maths and science. Most parents with more than one child can tell such stories.

Siblings are less like each other than would be expected just by chance.[15] The child picks a niche for itself, not on the basis of its own characteristics but on what its siblings have done. Such interplay between siblings probably accounts for some of the influences of birth order that we mentioned in Chapter 3. Once again individual differences emerge because children are active agents in their own development. Other things are also at work, of course. Parents treat their successive children differently – sometimes deliberately, sometimes unwittingly. They often have a more taut relationship with their first child than with later-born children, being more anxious and controlling.[16] They are usually more relaxed, positive and confident with their subsequent children and their preoccupation with every detail of their children's behaviour and appearance lessens. Many parents possess more family photographs of their first child than they do of later children.

For those who have them, siblings form an important component of the child's early environment. Consequently, the child's intellectual development depends in part on the number of siblings who are around to stimulate and to teach. The evidence from measurements of IQ suggests that the confluence of all these experiences enhances performance in IQ tests by a small, but nonetheless significant, amount. Having a younger sibling – to teach perhaps – also enhances the older child's performance in IQ tests.[17] The developmental benefits of having both older and younger siblings put the only child, who has neither, in a special position. The disadvantage of being an only child is small but measurable. Only children are widely believed to be spoilt and selfish.

Little Emperors?

If only children do grow up to be more selfish than others, then China (and by extension the rest of the world) will have to live with the consequences on a massive scale. Concerned about the economic and social impact of an exponential population growth, the Chinese government started to take firm action in the 1970s to forestall a catastrophe. A country of 1·2 billion people – one-fifth of the world's population – committed itself to a policy of one child per family. The one-child policy was implemented in various ways, using both threats and rewards.

Rewards included money and land allocations for farmers; punishments included fines, demotion, loss of employment and withdrawal of Communist Party membership. A second child is not allowed to enter the better schools in the community. By 1984, according to official estimates, twenty-four million couples had formally pledged to have no more than one child. Abortion is legal and women who already have one child are under social pressure to abort if they become pregnant again.

Some products of the single-child regime tend, it is claimed by outsiders, to be over-indulged, fat and self-absorbed. They are known as 'Little Emperors'. It has been suggested that these pampered children are a major educational and social problem in China and concern is expressed about what they might become. One school counsellor is reported as saying, 'I can't imagine what these egotistical, selfish and moody brats are going to be like when they grow up. I think this generation will be the most self-centred in Chinese history and will turn traditional Chinese ethics and morality on their heads.' Will China's determined and conscious attempt to curb its population growth have unforeseen effects on the collective psychological and physical characteristics of a whole country? At present, the scientific evidence is limited and equivocal.[18] The rest of the world can only stand back and watch anxiously what happens.

Rules for Changing the Rules

Learning is the most obvious way in which individuals interact with, and are changed by, their environment. Learning is entwined in the processes of human behavioural development, adapting individuals' behaviour to local conditions, enabling them to copy the behaviour of more experienced people, and fine-tuning preferences and actions that were inherited from previous generations. The multiple roles of learning in moulding a well-designed life almost certainly involve different ways of extracting knowledge about regularities in the environment.

Initially, at least some of these underlying rules must themselves develop without learning, since the process must start somehow – the biological equivalent of computer bootstrapping. Learning the rules for learning itself requires developmental regularity. The world is full of confusing information. Each individual must know what to store from a given experience and what to reject. In a well-designed life even the initial rules for learning must be adapted to the function they serve.

The best-known type of learning process was made famous by Ivan Pavlov a century ago. Pavlovian or classical conditioning, as it is called, allows the individual to predict what will be of real significance in the confusing world of sights, sounds and smells. Pavlov's famous experiment was to teach a dog to expect food by repeatedly alerting it with a buzzer before the food was presented. He then measured how much saliva the dog produced. As the dog was conditioned by the predictable association between the buzzer and the food, it came to produce saliva in response to the sound of the buzzer alone. As any dog owner knows, a hungry dog will do many other things once it detects cues that predict the arrival of food; it will go to the food bowl, whine, wag its tail, jump up and show all the familiar signs of expectation.

In Pavlovian conditioning, the sequence in which the events occur is crucial. If the buzzer is sounded *after* the presentation of food to a dog that has not yet been conditioned, it will not salivate or show any other expectant signs when the buzzer is subsequently sounded. The link in time between the action and the outcome is crucial. The rules for discovering causal relations between things make sense in terms of their utility because they map closely onto how the world works.

Another form of learning leads to the recognition of faces and places. The ability to distinguish between the vast array of objects, other people and scenes experienced in a lifetime is of inestimable value and happens simply as a result of exposure. The memory capacity is extraordinary. In one experiment people were shown up to 10,000 different slides projected onto a screen in quick succession. Days later they were able to recognise thousands of those images which they had previously seen. Some of the pictures, such as a dog smoking a pipe or a crashed aeroplane, were particularly vivid. In these cases the recall was even better and the subjects seemed to have a virtually limitless capacity to store them.[19]

Objects in the real world are rarely flat and their appearance depends on how they are viewed. A friend or relative is easily recognised from the front or the back, whether they are in the distance or close up. But they may not be so readily recognised if the photograph is taken from an odd angle such as from their feet. The recognisable features of a familiar person are fused together by the brain into a single category when these different views are seen in quick succession. Doubtless many other learning rules are also important in this case, such as lumping together different views seen in the same context. The point, though, is that time plays a different role in such perceptual learning than is usually the case in Pavlovian conditioning. The order in which different events are experienced is important when one event causes the other, but unimportant when the experiences are different views of the same object.

It seems likely that the initial rules for learning are themselves unlearned, universal and the products of Darwinian evolution. Does this mean that all human behaviour is predictable? The answer is emphatically 'No'. To understand why, consider a rule-governed game like chess. It is impossible to predict the course of a particular chess game from a knowledge of the game's rules. Chess players are constrained by the rules and the positions of the pieces in the game, but they are also instrumental in generating the positions to which they must subsequently respond. The range of possible games is enormous. Indeed, it is often unclear who is going to win until almost the end of the game. In a famous encounter between Robert Byrne and Bobby Fischer in 1963, Fischer seemed to have lost the game by the twenty-first move. He had just moved his queen and two grandmasters, who were providing a commentary for spectators, declared that Byrne would win. In fact, Byrne knew better and eventually resigned without making another move because he realised he would be mated within four moves. Underlying rules of clear adaptive significance can generate surface behaviour of enormous complexity, and inferring those underlying rules from watching many instances is difficult.

System and Synthesis

We have argued that order underlies even those learning processes that make people different from each other. Knowing something of the underlying regularities in development does bring an understanding of what happens to the child as it grows up. The ways in which learning is structured, for instance, affect how the child makes use of environmental contingencies and how the child classifies perceptual experience. Yet predicting precisely how an individual child will develop in the future from knowledge of the developmental rules for learning is no easier than predicting the course of a chess game. The rules influence the course of a life, but they do not determine it. Like chess players, children are active agents. They influence their environment and are in turn affected by what they have done. Furthermore, children's responses to new conditions will, like chess players' responses, be refined or embellished as they gather experience. Sometimes normal development of a particular ability requires input from the environment at a particular time; what happens next depends on the character of that input. The upshot is that, despite their underlying regularities, developmental processes seldom proceed in straight lines. Big changes in the environment may have no effect whatsoever, whereas some small changes have big effects. The only way to unravel this is to understand the developmental cooking processes.

5

Protean Instincts

Nature is often hidden, sometimes overcome, seldom extinguished.
Francis Bacon, 'Of Nature in Men', *Essays* (1625)

Original Matter

Proteus was the Old Man of the Sea, the shepherd of the sea mammals in Greek mythology. Knowing everything that had happened in the world, he was a universal historian. More importantly, he knew everything that would happen in the future and was therefore much in demand as a prophet. But he hated telling people what he knew. He would attempt to avoid divulging the future to those trying to consult him by slipping away from them, assuming many different shapes. Proteus came to be regarded by some as a symbol of the original matter from which the world was created.

Instinct is protean because it takes on many different forms, causing great confusion in the process. Instinct is also regarded by some as the basis from which all behaviour is created. In common usage instinct has various other connotations: they include 'biological' (as opposed to 'psychological'), 'hard-wired', 'congenital', 'endogenous', 'natural', 'inborn', 'constitutional', 'genetically determined' or simply 'genetic'.[1] These terms all carry similar theoretical connotations about the nature of behaviour. Instinctive behaviour patterns are, so it is said, inherited, internally motivated and adaptive. How can these ideas be reconciled with current understanding of the ways in which behaviour develops and, in particular, with the notion of developmental cooking? Suffice to say at this stage that the use of inherited, internally motivated and adaptive as defining characteristics presents problems of its own, making real behaviour patterns that match the concept of instinct all the more elusive. Charles Darwin recognised the problem in *The Origin of Species* in 1859:

> I will not attempt any definition of instinct. It would be easy to show that several distinct mental actions are commonly embraced by this term; but everyone understands what is meant when it is said that instinct impels the cuckoo to migrate and to lay her eggs in other birds' nests. An action, which we ourselves require experience to enable us to perform, when performed by an animal, more especially by a very young one, without experience, and when performed by many individuals in the same way, without their knowing for what purpose it is performed, is usually said to be instinctive. But I could show that none of these characters are universal. A little dose of judgment or reason . . . often comes into play, even with animals low in the scale of nature.

The reason why Darwin wisely refused to provide a comprehensive definition of instinct was because the concept has so many different dimensions to it. The same is true today. At their simplest, instincts may be nothing more than reflex reactions to external triggers, like the knee-jerk or the baby's sucking of a teat. In more complex forms, instincts are a series of movements co-ordinated into a system of behaviour that serves a particular end, such as locomotion or non-verbal communication. Later in his life Darwin published an article in which he described the behaviour of his first-born.[2] This article became the stimulus for the modern study of development in humans. Darwin wrote:

> During the first seven days various reflex actions, namely sneezing, hickuping, yawning, stretching, and of course sucking and screaming, were all performed by my infant. On the seventh day, I touched the naked sole of his foot with a bit of paper, and he jerked it away, curling at the same time his toes, like a much older child when tickled. The perfection of these reflex movements shows that the extreme imperfection of the voluntary ones is not due to a state of the muscles or of the co-ordinating centres, but to that of the seat of the will. At this time, though so early, it seemed clear to me that a warm soft hand applied to his face excited a wish to suck. This must be considered as a reflex or an instinctive action, for it is impossible to believe that experience and association with the touch of his mother's breast could so soon have come into play.

The behaviour of other animals provides plenty to marvel at. Consider the intricate way in which a spider builds a web. The spider explores a potential site, creates a frame for the web round the various attachment points, spins radials from the attachment points to the orb, spins more radials to the frame, then spins a spiral from the orb to the outside, and finally another spiral from the frame to the orb. It is an exquisite structure adjusted to the site in which it is built.[3]

At their most complex, instincts are thought to provide the basis by which the individual gathers particular types of information from the environment in the course of learning. The acquisition of language by humans is such a case. Children acquire words and the local rules of grammar from the adults around them, but the way they do so is thought to be shared by all humans. Therefore, the underlying process is believed to be inherited, internally motivated and adaptive.[4]

Instinct became a focus of intense interest during the middle of the twentieth century among biologists who were studying the behaviour of animals. Their interest was rekindled by the writings of a colourful and controversial figure called Konrad Lorenz.[5] He was born in 1903, the son of a rich and famous Austrian surgeon who had devised a way of correcting congenital deformities of the skeleton by 'bloodless surgery'. Lorenz's father tolerated the menagerie of animals kept by his son from when he was a small boy, but insisted that he train to become a doctor. Lorenz obeyed. But as soon as he was medically qualified, Lorenz went to Vienna and worked for a PhD in comparative anatomy, meanwhile maintaining his interest in the behaviour of the animals which he kept at the family home at Altenberg. His father remained tolerant of Lorenz's enthusiasms and supported him throughout the 1920s and 1930s when he had no paid post. During this period Lorenz's reputation as a scientist was growing and, by the 1930s, he was forging a new theory of instinctive behaviour with his Dutch friend Niko Tinbergen, with whom he and Karl von Frisch were to share the Nobel for physiology or medicine in 1973. Together, Konrad Lorenz and Niko Tinbergen became known as the founders of ethology – the biological study of behaviour.

Tinbergen came from an intellectual family which spawned two Nobel prize-winners; his brother won the prize for economics. Niko moved from Holland to Oxford in England after the Second World War. In 1951 he published a seminal book, *The Study of Instinct*, which set out the main findings and ideas of ethology. He particularly liked to study animals in their natural environments and he was a past master at conducting simple but elegant behavioural experiments in the field. Lorenz, on the other hand, preferred to keep animals in his home, where he could more readily observe them. He was struck by how behaviour patterns that had looked so appropriate in the natural worlds to which the animals had been adapted looked so odd when seen out of their normal context. A few days after hatching, a hand-reared duckling appears to spread oil over its down; it touches with its bill a pimple above its tail and then wipes its bill over its down. Yet this pimple, which becomes an oil-producing gland in the adult, is not yet functional and the duckling would normally be oiled by the feathers of its brooding mother. Observations such as this led Lorenz to conclude that behaviour patterns which were well adapted by evolution to the biological needs of the animal are qualitatively distinct from behaviour acquired through learning. Throughout his life, Lorenz fiercely contested what he regarded as the orthodoxies of American behaviourist psychology, with its almost exclusive emphasis on learning.[6]

Lorenz, with his academic training in comparative anatomy, believed that behavioural activities could be regarded like any physical structure or organ of the body. They had a regularity and consistency that related to the biological needs of the animal, and they differed markedly from one species to the next. But while Lorenz was a forceful advocate of the concept of instinct, he certainly did not deny the importance of learning. On the contrary, he gave great prominence to developmental processes by which animals formed their social and sexual preferences. He saw such learning processes as being under the control of what he referred to as the 'innate school marm'. She represented the highly regulated acquisition of information from the environment just when it is most adaptive for the animal to get it. Lorenz thought of instincts, whether they organised behaviour directly or were the mechanisms that changed behaviour through learning, as inherited neuronal structures which remained unmodified by the environment during development. Behaviour resulting from learning was seen as being separately organised in the brain from the instinctive elements.

Later in this chapter we shall take a critical look at Lorenz and Tinbergen's initial conception of instinct. We must stress, though, that their views were based on many compelling observations of animals' behaviour in the natural environment. These have been added to by a great wealth of evidence in subsequent years. And many examples of courtship, defensive behaviour, specialised feeding methods, communication and much else are familiar to a wide audience from exquisite television films. Some behaviour patterns are highly stereotyped in their form, and are stable across a wide range of environmental conditions. The web-making abilities of spiders are like this. Complex and co-ordinated behaviour patterns may also develop without practice. Birds, for example, can usually fly at their first attempt, without any apparently relevant prior experience. In one experiment, young pigeons were reared in narrow boxes that physically prevented them from moving their wings after hatching. They were then released at the age at which pigeons normally start to fly. Despite having had no prior opportunities to move their wings, the pigeons were immediately able to fly when released, doing so almost as well as the pigeons which had not been constrained.[7] In a similar way, European garden warblers which have been hand-reared in cages nevertheless become restless and attempt to fly south in the autumn – the time when they would normally migrate southwards. The warblers continue to be restless in their cages for about a couple of months, the time taken to fly from Europe to their wintering grounds in Africa. The following spring, they attempt to fly north again. This migratory response occurs despite the fact that the birds have been reared in social isolation, with no opportunities to learn when to fly, where to fly or for how long.[8]

The Universal Smile

'Be a good animal, true to your instincts,' wrote D.H. Lawrence. But instances of apparently instinctive or innate behaviour patterns are not confined to so-called simpler animals, as Darwin himself noted. Human babies who have been born blind, and consequently never able to see a human face, nevertheless start to smile at around five weeks – the same age as normal babies.[9] Babies do not have to see other people smile in order to smile themselves. Just after birth, sighted human babies gaze preferentially at head-like shapes that have the eyes and mouth in the right places. Invert these images of heads, or jumble up the features, and the newborn babies respond much less strongly to them.[10]

The brains of humans and other primates contain specialised networks of neurons that respond specifically to faces and facial expressions. Two quite separate and distinct regions of the human brain seem to be involved in the processes of, first, encoding new memories for faces, and then recognising those faces at some later date.[11] Brain-scanning has revealed that the left prefrontal cortex of the brain is activated during the encoding of faces, whereas the right prefrontal cortex is activated during later recognition. When adults lose these special face-recognition networks as the result of a stroke, tumour or surgery, they are rendered unable to recognise faces. People with this distressing condition, which is known as prosopagnosia, cannot even recognise their own partners; when meeting in a crowd, the partner must wear something conspicuous.

Human facial expressions have universal characteristics that transcend cultures. The emotions of disgust, fear, anger and pleasure are read off the face with ease in any part of the world. Towards the end of his life Charles Darwin wrote *The Expression of the Emotions in Man and Animals*, a book that provided the stimulus for observational studies of animal and human behaviour which have continued into modern times. Darwin concluded,

'That the chief expressive actions, exhibited by man and by the lower animals, are now innate or inherited – that is, have not been learnt by the individual, – is admitted by every one.' Darwin's descriptions of suffering, anxiety, grief, joy, love, sulkiness, anger, disgust, surprise, fear and much else are models of acute observation. He would show to friends and colleagues pictures of a man expressing various emotions and ask them, without further prompting, to describe the emotions. In one case a picture of an old man with raised eyebrows and open mouth was shown to twenty-four people without a word of explanation, and only one did not understand exactly what was being expressed. Darwin continued, 'A second person answered terror, which is not far wrong; some of the others, however, added to the words surprise or astonishment, the epithets, woful, painful, or disgusted.' His extensive correspondence with travellers and missionaries convinced him that humans from all round the globe expressed the same emotions in the same ways.

The study of facial expressions which Darwin started was resumed vigorously nearly a century later. One of Konrad Lorenz's former students, Irenäus Eibl-Eibesfeldt, visited many remote aboriginal people who had had little or no contact with the outside world. He built up an enormous archive of photographic records of human expressions in different cultures at different stages of economic development. The similarities in, for example, the appearance of the smile or the raised eyebrows are striking. The cross-cultural agreement in the interpretation of complex facial expressions is also remarkable. People agree about which emotions are being expressed, even when an expression denotes two or more different emotions of different intensities. They also agree about which emotion is the more intense, such as which of two angry people seems the more angry.[12]

Certain aspects of human behavioural development stand out as solid, unchanging structures on the shifting sands of cultural differences and the unique contingencies of any one person's life.

Despite the host of genetic and environmental influences that contribute to behavioural differences between individuals, all members of the same species are remarkably similar to each other in many aspects of their behaviour – at least, when compared with members of other species. All humans have the capacity to acquire language, and the vast majority do. With few exceptions, humans pass the same developmental milestones as they grow up. Most children have started to walk by about eighteen months after birth, have started to talk by around two years, and go on to reach sexual maturity before their late teens. Individual differences among humans seem small when any human is compared with any chimpanzee. This is not surprising because if different individuals of the same species were built to fundamentally different biological designs then the shuffling of genes which occurs during sexual reproduction would seldom produce a body that worked. The evidence suggesting that the development of behaviour had many of the same characteristics as the development of the body led to popular arguments about the supposed human instincts for holding territory, cheating and, in Lorenz's hands, aggression.

The Stink of Instinct

By the mid-twentieth century, those studying behaviour, particularly in the United States and the Soviet Union, found the late nineteenth-century hereditarian notions offensive. In America the ideology of individualism suggested that everybody could be instrumental in their own route to personal success, while in Russia the ideology imposed by Lenin and then Stalin suggested that everybody could be shaped to serve the interests of the state. Behaviourist psychology in the United States and Pavlovian psychology in the Soviet Union conformed with their respective political ideologies through the emphasis they placed on learning in the development of behaviour. A new reductionism was born, namely environmental determinism. All this academic activity led to the decline in influence of the whole idea of instinct in human psychology.

Despite all the empirical evidence that some elements of behaviour can develop without opportunities for learning, the ethologists' notion of instinct attracted strong criticism in the 1950s from a group of American comparative psychologists who studied animal behaviour. A major figure in this camp was Ted Schneirla, who worked at the American Museum of Natural History in New York and was famous for his work on the behaviour of ants. The attacks on instinct were in part motivated by Lorenz's pre-war acceptance of the ideology of the Third Reich. When the Nazis came to power, Lorenz had swum with the tide and in 1940 wrote a shocking article that dogged him for the rest of his life. He deplored the effects of domestication on animal species and thought (without any evidence) that humans were becoming victims of their own self-domestication. Having got 'our best individuals to define the type-model of our people', the unfortunates who deviated markedly from such a model should, he suggested, be prevented from breeding. Lorenz's wish to rid humanity of what he regarded as impurity matched only too well Hitler's appalling dream.[5]

Whatever the original motivation for the intellectual assault on Lorenz's ideas about instinct, the attacks hit home, and the critics laid out a quite different agenda for studying behavioural development. A leading protagonist in the debate was Danny Lehrman, a brilliant, ebullient and enormously articulate man who later founded the Institute of Animal Behavior at Rutgers University in New Jersey.[13] Lehrman's critique of Lorenz was accepted by Niko Tinbergen and shaped the thinking of most English-speaking ethologists. Indeed, the ideas in this book about the cooking processes of development are firmly rooted in the intellectual tradition that emerged from Lehrman and his colleagues. An important American psychobiologist, Frank Beach, referred to this change in thinking as 'the descent of instinct',[14] or, in private, as 'taking the stink out of instinct'. Lehrman made the point this way:

> The problem of development is the problem of the development of new structures and activity patterns from the resolution of the interaction of existing ones, within the organism and its internal environment, and between the organism and its outer environment. At any stage of development, the new features emerge from the interactions within the current stage and between the current stage and the environment. The interaction out of which the organism develops is not one, as is often said, between heredity and environment. It is between organism and environment! And the organism is different at each stage of its development.[15]

The stink (as some would see it) of instinct has resurfaced strongly in the late twentieth century, in the writings of sociobiologists and evolutionary psychologists. The debate has been confused because the term 'instinct' means remarkably different things to different people.[16] To some, 'instinct' means a distinctly organised system of behaviour patterns, such as that involved in searching for and consuming food. The different modules of behaviour have been likened to the various tools found on a Swiss Army knife.[17] For others, an instinct is simply behaviour that is not learned. Instinct has also been used as a label for behaviour that is present at birth (the strict meaning of 'innate') or, like sexual behaviour, patterns that develop in full form at a particular stage in the life cycle. Another connotation of instinct is that once such behaviour has developed it does not change. Instinct has also been portrayed as behaviour that develops before it serves any biological function, like some aspects of sexual behaviour. Instinct is often seen as the product of Darwinian evolution so that, over many generations, the behaviour was adapted for its present use. Instinctive behaviour is supposedly shared by all members of the species (or at least by members of the same sex and age). Confusingly, it has also been used to refer to a behavioural difference between individuals caused by a genetic difference – so instincts are both universal and part of individual differences. The overall effect is, to say the least, muddling and brings to mind Humpty Dumpty's conversation with Alice in *Through the Looking Glass*:

> 'When *I* use a word,' Humpty Dumpty said in rather a scornful tone, 'it means just what I choose it to mean – neither more or less.'
>
> 'The question is,' said Alice, 'whether you *can* make words mean so many different things.'
>
> 'The question is,' said Humpty Dumpty, 'which is to be master – that's all.'

Some real examples of behaviour can be found to which most of the defining characteristics of instinct seem to apply, such as the ways in which mice, rats and guinea pigs clean their own fur. The duration of the elliptical stroke with the two forepaws which the rodent uses to clean its face is proportional to the size of the species; the bigger the species the longer the stroking movement takes. This is not simply a matter of physics. The bigger-bodied animals are not slower in their grooming movements simply because their limbs are heavier; a baby rat grooms at exactly the same rate as an adult rat even though it is a tenth of the size.[18] Moreover, young rodents perform these grooming movements at an age when their mother cleans them and before the behaviour is needed for cleaning their bodies. Rodent grooming is, in other words, a species-typical, stereotyped system of behaviour that develops before it is of any use to the individual. It has the main defining characteristics of instinct. But real instances of behaviour are not always as clear as this.

A Little Dose of Judgement

Many of the theoretical implications of the concept of instinct are difficult to test in practice. Take the definition of instinctive as being unlearned, for instance. To establish experimentally that a particular type of behaviour is not learned requires the complete exclusion of all opportunities for learning. This is harder than it sounds. For a start, it is difficult to draw a clear distinction between experiences that have specific effects on the detailed characteristics of a fully developed behaviour pattern and environmental influences that have more general effects on the organism, such as nutrition or stress. Experiences vary in the specificity of their effects.

Even if all obvious opportunities for learning a particular behaviour pattern are excluded, a major problem remains. This is because animals, like humans, are good at generalising from one type of experience to another. It is therefore difficult to know whether an individual has transferred the effects of one kind of experience to what superficially looks like a quite different aspect of their behaviour. For example, if somebody draws a letter of the alphabet on your hand while your eyes are shut you should still be able to visualise the letter, even though you have not seen it. In doing this you will have demonstrated a phenomenon called cross-modal matching. A striking instance of cross-modal matching has been found in rhesus monkeys. Researchers trained monkeys to distinguish between tasty and obnoxious biscuits in the dark. The obnoxious biscuits, which contained sand and bitter-tasting quinine, differed in shape from the tasty biscuits. The monkeys quickly learned to select the right-shaped biscuits. When they were subsequently tested in the light, the monkeys immediately reached for the nice biscuits, even though they had never seen them before. They had transferred the knowledge they had acquired from a purely tactile experience – touching the biscuits in the dark – and used it to make a visual choice.[19]

Another pitfall in the quest for instinct is that the developing individual cannot be isolated from itself, and some of its own actions may provide crucial experience that shapes its subsequent behaviour. After they hatch, ducklings exhibit an immediate preference for the maternal calls of their own species. Experiments by the American psychobiologist Gilbert Gottlieb showed that the ducklings' preferences are affected by their hearing their own vocalisations in the egg before hatching. In other words, their 'instinctive' preference is influenced by sensory stimulation which they generated themselves. Gottlieb was able to demonstrate this by cutting a hole in the egg and operating on the unhatched ducklings, thereby making it impossible for them to produce sounds. These silent birds were less able to distinguish the maternal calls of their own species from those of others. However, if they were played tape-recordings of duckling calls while they were in the egg, the preference for their own species' maternal call emerged as normal.[20]

A formidable obstacle to proving that a behaviour pattern is not learned is the capacity that animals have to acquire the necessary experience in more than one way. When scientists attempt to isolate an animal from one particular form of experience that is thought necessary for development, the behaviour pattern may nonetheless develop by an alternative route. Cats, for example, can acquire and improve their adult predatory skills via a number of different developmental routes: by practising catching live prey when young, by playing at catching prey when young, by watching their mother catch live prey, by playing with their siblings, or by practising when adult. Hence a kitten deprived of, say, opportunities for play may still develop into a competent adult predator, but by a different developmental route.[21]

The demonstrations of cross-modal matching, the impact of self-stimulation on the young animal and the use of different developmental routes to the same end point all sound notes of warning. It is not as easy as it might seem to demonstrate that a behaviour pattern has not been shaped by some form of experience. The various characteristics of instinct do not always hang together so closely as they do in the example of rodent grooming. A central aspect of Lorenz's concept of instinct that unravelled on further inspection was the belief that learning does not influence such behaviour patterns once they have developed. Many cases of apparently unlearned behaviour patterns are subsequently modified by learning after they have been used for the first time. A newly hatched laughing gull chick will immediately peck at its parent's bill to initiate feeding, just as, in the laboratory, it will peck at a model of an adult's bill. At first sight this behaviour pattern seems to be unlearned; the chick has previously been inside the egg and therefore isolated from any relevant experience, so it cannot have learned the pecking response. However, as the chick profits from its experience after hatching, the accuracy of its pecking improves and the kinds of model bill-like objects which elicit the pecking response become increasingly restricted to what the chick has seen. Here, then, is a behavioural response that is present at birth, species-typical, adaptive and unlearned, but nonetheless modified by the individual's subsequent experience.[22]

Essentially the same is true for the 'innate' smiling of a blind human baby. Smiling behaviour emerges soon after birth in blind babies, as it does in sighted babies. But sighted people subsequently learn to modify their smiles according to their experience, producing subtly different smiles that are characteristic of their particular culture. The blind child, lacking the visual interaction with its mother, becomes less responsive and less varied in its facial expression.[23] The fact that a blind baby starts to smile in the same way as a sighted baby does not mean that learning has no bearing on the later development of social smiling. Experience can and does modify what started out as apparently unlearned behaviour. Conversely, some learned behaviour patterns are developmentally stable and virtually immune to subsequent modification. We shall describe in Chapter 8 how the songs of some birds are learned early in life; in some species, though, these learned songs are extremely resistant to change once they have been acquired.[24]

The idea that one meaning of instinct, 'unlearned', is synonymous with another, namely 'adapted through evolution', also fails to stand up to scrutiny. The development of a behaviour pattern that has been adapted for a particular biological function during the course of the species' evolutionary history may nonetheless involve learning during the individual's lifespan. For example, the strong social attachment that young birds and mammals form to their mothers is clearly adaptive and has presumably evolved by Darwinian evolution. And yet the attachment process requires the young animal to learn the individual distinguishing features of its mother.

Yet another area in which the different elements of instinct fall apart is in the role of learning in the inheritance of behaviour across generations. Consider, for example, the ability of birds such as titmice to peck open the tops of the milk bottles that used to be delivered each morning to the doors of most British homes. The birds' behaviour was clearly adaptive, because they gained access to the cream that rose to the tops of these bottles. Exploiting a valuable source of fatty food undoubtedly increased the individual bird's chances of surviving the winter and breeding the following spring. However, the bottle-opening behaviour pattern was transmitted from one generation to the next by means of social learning. The basic tearing movements used in penetrating the bottle top are also used in normal foraging behaviour and were probably inherited without learning. But the trick of applying these movements to opening milk bottles was acquired by each individual bird through watching other birds do it successfully – that is, by social learning.[25] (How the original birds first discovered the trick is another matter.)

In short, many behaviour patterns have some, but not all, of the defining characteristics of instinct, and the unitary concept starts to break down under closer scrutiny. The various theoretical connotations of instinct – namely that it is unlearned, caused by a genetic difference, adapted over the course of evolution, unchanged throughout the life-span, shared by all members of the species, and so on – are not merely different ways of describing the same thing.

Managing the Machine

Within an individual, different systems of behaviour are variable, both in their development and in their organisation. Some insight into why this should be may be obtained by looking at machines. Tailoring a system to a specified job while building in flexibility is a problem that human designers of machines must face again and again. Robots with even simple forms of regulation do things that look remarkably lifelike. Similarly, in a game like chess simple rules can generate games of great complexity – even when played by machines using straightforward software. Nevertheless, the difficult challenge for the designers of such machines is to beat the creativity, flair and imagination of a chess grandmaster.

IBM rose to the chess challenge and started its Deep Blue project in 1989. The Deep Blue computer relied on massive parallel arrays with dedicated hardware and software. It had 256 chess-specific processor chips operating in tandem, each capable of analysing up to three million chess moves every second. The whole array could process about 250 million moves each second, or 50–100 billion moves in the three minutes allotted for each move. It was also equipped with an enormous database of grandmaster games played in the past century. In the initial stages of the project no attempt was made to mimic human thought. Without any 'psychology' to mess things up, the machine would never get tired or make a silly mistake. It would instead depend for its success on raw computing power and its enormous memory. In one second Deep Blue could search ahead through several hundred million moves, while its human opponent, the Russian grandmaster and one-time world champion Gary Kasparov, could analyse only one or two. Kasparov himself admitted that quantity sometimes becomes quality. But he had the compensatory benefits of intuition, judgement and experience. Compared to computers, humans calculate slowly, but are good at recognising patterns. Chess grandmasters are much better than novice players at remembering patterns of pieces from real games, but no better at remembering arbitrary patterns. Experience helps them to remember patterns that have meaning and link these with the sequences of moves that have the best pay-off in the long run. The surprising consequence is that humans see traps that lie beyond the search horizon of even an exceedingly fast computer.

In 1996 Kasparov played Deep Blue in a six-game match. Kasparov lost the first game, but then put his human skill to good effect and went on to win the match. He was able to do this because he could adapt his strategy in response to what he discovered were weaknesses in his machine opponent. Deep Blue, on the other hand, could not respond to the overview of its human opponent. IBM rose to the new challenge. Deeper Blue, their 1997 successor to Deep Blue, was faster and smarter. In particular, it could modify its basic strategy between tournaments in response to the playing style of its human adversary. This time the machine managed, albeit with some difficulty, to win the next match against Kasparov.

The chess matches emphasised how important adaptability is in such competitions. In more practical uses, such as the control of traffic flow by co-ordinating the switching of traffic lights or regulating speed limits, a capacity to adapt to new situations is desirable. Faced with novelty, such systems must not change everything, however. If they did, they would quickly collapse into chaotic malfunctioning. So, as with animals, buffering some aspects of the computer's capacities from change is crucial. These essential capacities must continue to function in the same way despite radical changes in input. This brings us back to the problems of development and the organisation of behaviour.

Impressive though IBM's Deeper Blue computer was, it was dedicated to one complicated but narrowly defined task – playing chess. Gary Kasparov may have met his match on the chessboard, but he was able to do a great many other complicated things of which Deeper Blue was incapable. He could make decisions about chess matches, and holidays which he would take years into the future. He could run a complicated social life and allocate time (we presume) to all his main biological appetites, none of which were shared with Deeper Blue. He could feel moved by patriotism or spiritual feelings. He could write books and enjoy music. From time to time, he doubtless reflected on his life and his character. Kasparov, like every other human and every other animal, had many strands to his life. The systems that are involved in running each of these aspects sometimes seem to be semi-autonomous, usually functioning smoothly together, but occasionally coming into conflict. Humans feel the conflict most strongly, perhaps, in times of war, when their craving for leadership and their identification with their own group, tribe or nation conflicts with their peacetime commitments and pleasures and, indeed, their perception of their own self-preservation. But everybody feels the pull, on most days of even the most routine life, between incompatible activities. You can't eat and sleep at the same time; you can't have a bath and take a walk simultaneously.

Much of animal and human behaviour and physiology operates on the basis that considerable autonomy has seemingly been designed into each behavioural system or organ. Interaction necessarily occurs between them to prevent total breakdown when the different parts pull in different directions. A problem of great interest to engineers has been how far machines should emulate biology, using specialised modules like those in the brain that are dedicated to particular jobs such as recognising faces. When the engineer does so, how far should the modules be built into separately organised systems, each competing for time when they cannot operate simultaneously? Should a 'boss' allocate priority where it is impossible for two activities to occur at the same time? Or should a decision to express a particular form of behaviour depend on weighting the needs of competing systems? Designers of intelligent machines often opt for distributed control, known as heterarchy (as opposed to hierarchy), because of the efficiency it brings.

The solutions to the problems of running a smart machine, or an individual life, are also found in the management of human organisations. In contrast with traditional hierarchical bureaucracies, modern companies and organisations increasingly tend to have structures and organisational cultures that focus on tasks or projects. The emphasis is on getting the job done efficiently and this is achieved by bringing together groups of people with the right knowledge and skills. Expertise and teamwork are what counts, rather than formal status. The organisational structure tends to be a matrix of project teams rather than a traditional top-down hierarchy. Such management relies on great flexibility and considerable autonomy for each part of the organisation, with exchange of information and competition occurring when the well-being of the whole demands it. The central control over day-to-day work is minimal and the ways in which each team is set up depends on the need.

These issues, which crop up in many areas of human life, reflect the need to find efficient solutions to complex organisational problems. The particular solution will depend on the priority given to each activity and the priorities will vary from case to case. How do animals achieve comparable solutions in the development and integration of their behaviour? The organised systems of behaviour, which have been called instincts by some, must reflect the function for which they are seemingly designed.

A Well-Designed Life

'Biology is the study of complicated things that give the appearance of having been designed for a purpose,' wrote Richard Dawkins in *The Blind Watchmaker*. Dawkins took the image of the watchmaker from an argument developed by William Paley in the early nineteenth century. 'It is the suitableness of these parts to one another; first, in the succession and order in which they act; and, secondly, with a view to the effect finally produced,' wrote Paley about the reaction of someone who contemplates the construction of a well-designed object.

The perception that behaviour is designed springs from the relations between the behaviour, the circumstances in which it is expressed and the resulting consequences. The closeness of the perceived match between the tool and the job for which it is required is relative. In human design, the best that one person can do will be exceeded by somebody with superior technology. If you were on a picnic with a bottle of wine but no corkscrew, one of your companions might use a strong stick to push the cork into the bottle. If you had never seen this done before, you might be impressed by the selection of a rigid tool small enough to get inside the neck of the bottle. The tool would be an adaptation of a kind. Tools that are better adapted to the job of removing corks from wine bottles are available, of course, and an astonishing array of devices have been invented. One ingenious solution involved a pump and a hollow needle with a hole near the pointed end; the needle was pushed through the cork and air was pumped into the bottle, forcing the cork out. Sometimes, however, the bottle exploded and this tool quickly became extinct. As with human tools, what is perceived as good biological design may be superseded by an even better design, or the same solution may be achieved in different ways.

Among those who spin stories about biological design, a favourite figure of fun is an American artist called Gerald Thayer. He argued that the purpose of the plumage of all birds is to make detection by enemies difficult. Some of the undoubtedly beautiful illustrations were convincing examples of the principles of camouflage. However, among other celebrated examples, such as pink flamingos concealed in front of the pink evening sky, was a painting of a peacock with its resplendent tail stretched flat and matching the surrounding leaves and grass. The function of the tail was to make the bird difficult to see![26] Ludicrous attributions of function to biological structures and behaviour have been likened to Rudyard Kipling's *Just So Stories* of how, for example, the leopard got his spots. However, the teasing is not wholly justified. Stories about current function are not about how the leopard got his spots, but what the spots do for the leopard now. That question is testable by observation and experiment.

Not every speculation about the current use of a behaviour pattern is equally acceptable. Both logic and factual knowledge can be used to decide between competing claims. Superficially attractive ideas are quickly discarded when the animal is studied in its natural environment. The peacock raises his enormous tail in the presence of females and he moults the cumbersome feathers as soon as the spring breeding season ends. If Thayer had been correct about the tail feathers being used as camouflage, the peacock should never raise them conspicuously and he should keep them all year round.

For animals, the ultimate arbiter of priority in organising their own behaviour is reproductive success and the consequences are sometimes astonishing – at least when judged from a human perspective. The male emperor penguin brooding his mate's egg over the Antarctic winter cannot be relieved by his mate because the growth of the ice shelf puts the sea and food beyond reach. So, in the interests of producing an offspring, he fasts for months – a feat any human would find impossible. Other potential solutions to this problem, such as shorter stints of brooding and trekking repeatedly across the ice shelf during the winter, were presumably less efficient. The penguins which fasted all winter were the ones that had the best design. Examples like this emphasise how dependent is the organisation of behaviour on the ecology of the species. A corollary is that the particular way in which a given system develops will also depend on ecology. Differences in the cooking processes of development are to be expected.

Flexible Structure

Behaviour cannot be neatly divided up into two distinct types: learned and instinctive. Nevertheless, Lorenz's belief that behaviour has something in common with the organs of the body does have substance. The developmental progression from a single cell to an integrated body of billions of cells, combining to produce coherent behaviour, is astonishingly orderly. Just as animals grow kidneys with a specialised biological function, adapted to the conditions in which they live, so they perform elaborate and adaptive behaviour patterns without any previous opportunities for learning or practice. Particular behaviour patterns are like body organs in serving particular biological functions; their structure is often likely to have been adapted to its present use by Darwinian evolution and depends on the ecology of the animal; and they develop in a highly co-ordinated and systematic way.

From the standpoint of design, systems of behaviour that serve different biological functions, such as cleaning the body or finding food, should not be expected to develop in the same way. In particular, the role of experience is likely to vary considerably from one type of behaviour to another. In predatory species such as cats, cleaning the body is not generally something that needs special skills tailored to local conditions, whereas capturing fast-moving prey requires considerable learning and practice to be successful. The osprey snatching trout from the water does not develop that ability overnight. Animals that rely upon highly sophisticated predatory skills, such as birds of prey, suffer high mortality when young, and those that survive are often unable to breed for years because they have to hone their skills before they can capture enough prey to feed offspring. In such cases, a combination of different developmental processes is required in order to generate the highly tuned skills seen in the adult.

Bringing together the ideas of design with those of development means that the adaptiveness of the developmental processes must be examined. We shall consider some examples in Chapter 6. As for the term 'instinct', it is a hold-all – a kitchen drawer if you will – containing a heterogeneous collection of implements with different uses. Or, to return to Proteus, the Old Man of the Sea, the various systems of behaviour take on different forms. Each should be described as it is found, and its defining characteristics are often different. Evidence that a behaviour pattern serves a current biological function does not constitute evidence that the behaviour pattern is unlearned, since crucial information needed for its development may come from different sources. The notion of design has helped to make sense of what otherwise looks like a confused jumble of implements in the kitchen drawer.

The flexibility in the ways that behaviour develops highlights a worry that haunts discussions of this subject. If it were supposed that humans responded like robots to the imperatives of survival and reproduction, then what is to be said about choice, freedom of will and personal responsibility? Shakespeare's answer is given by Coriolanus:

I'll never
Be such a gosling to obey instinct, but stand
As if a man were author of himself
And knew no other kin.

Humans sometimes cannot help themselves. More often, though, they can and should. We shall explore the role of conscious choice in development in Chapter 7.

6

Alternative Lives

> One is born manifold and one dies single.
>
> Paul Valéry, cited by Simone de Beauvoir, *All Said and Done* (1974)

The Developmental Jukebox

In one of his short stories, Ernest Hemingway described a traveller who was walking through a landscape that had been devastated by fire:

> Nick sat down against the charred stump and smoked a cigarette . . . As he smoked, his legs stretched out in front of him, he noticed a grasshopper walk along the ground and onto his woolen sock. The grasshopper was black. As he walked along the road, climbing, he had started many grasshoppers from the dust. They were all black.

Hemingway got it right. After a fire on the high grassland planes of East Africa, for example, the young grasshoppers are indeed black instead of being the normal pale yellowish-green. Something has switched the course of their development onto a different track. The grasshopper's colour makes a big difference to the risk that it will be spotted and eaten by a bird, and the scorched grassland may remain black for many months after a fire. So matching its body colour to the blackened background is important for its survival. The developmental mechanism for making this switch in body colour is automatic and depends on the amount of light reflected from the ground.[1] If the young grasshoppers are placed on black paper they are black when they moult to the next stage. But if they are placed on pale paper the moulting grasshoppers are the normal green colour. The grasshoppers actively select habitats with colours which match their own. If the colour of the background changes they can also change their colour at the next moult to match the background, but they are committed to a colour once they reach adulthood.

Turtles and crocodiles and some other reptiles commit themselves early in life to developing along one of two different developmental tracks and, like grasshoppers, they do so in response to a feature of their environment.[2] Each individual starts life with the capacity to become either a male or a female. The outcome depends on environmental temperature during the middle third of embryonic development.[3] If the eggs from which they hatch are buried in sand below 30°C, the young turtles become males. If, however, the eggs are incubated at above 30°C, they become females. Temperatures below 30°C activate genes responsible for the production of male sex hormones and male sex hormone receptors. If the incubation temperature is above 30°C, a different set of genes is activated, producing female hormones and receptors instead.[4] In alligators the sex determination works the other way around, so that eggs incubated at higher temperatures produce males. (In humans and other mammals, by contrast, the sex of each individual is determined genetically at conception; if it inherits only one X sex chromosome it becomes male.)

Each grasshopper and turtle starts life with the capacity to play two distinctly different developmental tunes – green or black, male or female. A particular feature of the environment then selects which of those tunes the individual will play during its life. And, once committed, the individual cannot switch to the other tune. Once black as an adult, the grasshopper cannot subsequently change its colour to green, just as a male turtle cannot transform itself into a female.

The broad pattern of an individual's social and sexual behaviour may also be determined early in life, with the individual developing along one of two or more qualitatively different tracks. Many examples are found in the animal kingdom. The caste of a female social insect is determined by her nutrition early in life. The main egg producer of an ant colony, the queen, is part of a teeming nest in which some of her sisters care for her offspring, others forage, yet others clean or mend the nest, and finally other sisters specialise in guarding it.[5] Locusts may or may not become migratory, depending on crowding; when the numbers living in a given area build up, their offspring develop musculature and behaviour suitable for long flights and then the whole swarm moves off.[6] Vole pups born in the autumn have much thicker coats than those born in spring; the cue to produce a thicker coat is provided by the mother before birth. The value of preparing in this way for colder weather is obvious.[7]

The sexual behaviour of some primates can also develop along two or more distinctly different tracks. An adult male gelada baboon, for example, will typically defend and breed with a harem of females. After a relatively brief but active reproductive life, he is displaced by another male and never breeds again. To position himself so that he can acquire and defend a harem, the male must grow rapidly at puberty. He develops the distinctive golden mane of a male in his prime and becomes almost twice the size of the females. However, when many such males are present in the social group, an adolescent male may adopt a distinctly different style of reproductive behaviour. He does not develop a mane or undergo a growth spurt. Instead, he remains similar in appearance and size to the females. These small males hang around the big males' harems, sneakily mating with a female when the harem-holder is not paying attention. Since the small, sneaky male never has to fight for females, he is likely to have a longer, if less intense, reproductive life. If he lasts long enough he may even do better in terms of siring offspring than a male who pursues the alternative route of growing large and holding a harem. These two different modes of breeding behaviour represent two distinctly different developmental routes, and each male baboon must commit himself to one or other of them before puberty.[8]

All these examples illustrate a surprising aspect of development which has intriguing implications for humans. In each case, the individual animal starts its life with the capacity to develop in a number of distinctly different ways. Like a jukebox, the individual has the potential to play a number of different developmental tunes. But during the course of its life it plays only one tune. The particular developmental tune it does play is selected by a feature of the environment in which the individual is growing up – whether it be the colour of the ground, the temperature of the sand, the type of food, or the presence of other males. Furthermore, the particular tune that is selected from the developmental jukebox is adapted to the conditions in which it is played.

What are the implications for humans of the concept of the developmental jukebox? Is it the case that people, like grasshoppers or baboons, are conceived with the capacity to play a number of qualitatively different developmental tunes – in other words, to live alternative lives? Each of us started life with the capacity to live many different lives, but each of us lives only one. In one sense individual humans are obviously bathed in the values of their own particular culture and become committed by their early experience to behaving in one of many possible ways. Differences in early linguistic experience, for example, have obvious and long-lasting effects. By the end of a typical high school education, a young American will probably know about 50,000 different words.[9] The words are different from those used by a Russian of the same age. In general, individual humans imbibe the particular characteristics of their culture by learning (often unwittingly) from older people.

When environmental conditions select a particular developmental route in animals, the mechanisms involved are likely to be different; learning may not enter into the picture at all. Is it possible that some aspects of human development are triggered by the environment, as though the individual were a jukebox? Was each of us conceived with the capacity to develop along a number of different tracks – to live a number of distinctly different sorts of life? And does the environment select the particular developmental track that each of us follows?

Forecasting the Weather

An individual's learning ability and memory may be subtly affected by its mother's nutritional state during pregnancy. If rat mothers have a substance called choline added to their diet during a particular period in the later stages of pregnancy, their pups have better long-term memories when they become adults.[10] Choline is needed by the developing pups to construct particular brain circuitry involved in memory. Of course, a lack of choline in the mother's diet might simply have a pathological effect on her pups which the supplement corrects. But if long-term memory is important for survival, why should it be so dependent on the vagaries of maternal nutrition?

A more interesting, but speculative, hypothesis suggests itself. Perhaps the unborn pups receive a 'weather forecast' from their mother that prepares them for the type of world in which they will have to live. If a shortage of choline in their mother's diet predicts a harsh world – and therefore a short life – for the developing pups, it may be to their advantage to cut their losses and have a short sprint to reproduction. If developing a superior long-term memory is costly then this may be one of the casualties. Conversely, if the mother's nutritional state predicts a friendlier world and a longer life, then a good long-term memory may be a luxury that the developing pups can afford to invest in. What about humans?

A series of studies led by the epidemiologist David Barker, which assessed people across their entire lifespan from birth to death, has lent strength to the suggestion that human development may also involve environmental cues that prepare the individual for a particular sort of environment.[11] The work was based in part on a large sample of men born in the English county of Hertfordshire between 1911 and 1930. The researchers found that those men who had had the lowest body weights at birth and at one year of age were most likely to die from cardiovascular disease later in life. The heaviest babies faced a subsequent risk of dying from cardiovascular disease that was only half the average for the group as a whole, whereas the risk for the smallest babies was 50 per cent above average (in other words, three times greater than that for the largest babies). Individuals who had been small babies were also more likely to suffer from diseases such as diabetes and stroke in adulthood.

How could a link have arisen between an individual's birth weight and his physical health decades later? The evidence pointed to a connection with the mother's nutritional state: women with poor diets during pregnancy had smaller placentas, and forty years later their offspring had higher blood pressure (a risk factor for cardiovascular disease and stroke). But the links with maternal nutrition went much further back than pregnancy. Measurements of the mothers' pelvises revealed that those who had a flat, bony pelvis tended to give birth to small babies. These small babies, after they had grown up, were much more likely as adults to die from stroke. The implication was that poor nutrition during their mother's childhood affected the growth of her pelvis which, in turn, curtailed the growth of her offspring during pregnancy which, in turn, increased her offspring's risk of stroke and cardiovascular disease in adulthood.

Some surprising connections also emerged between early physical development and adult fingerprint patterns.[12] Individuals who had been thinner and shorter at birth were found to have more whorl patterns on the fingers of their right hand than individuals who had been fatter and longer as babies. It appears that whorls on the fingertips are indelible markers of reduced foetal growth, which is linked to raised blood pressure in adulthood and consequently an increased risk of cardiovascular disease.

Poor maternal physique and health are associated with reduced foetal growth, with consequences for the offspring's later health. The question arises, then, as to whether these connections make sense in adaptive terms. Could it be that, in bad conditions, the pregnant woman unwittingly signals to her unborn baby that the environment which her child is about enter is likely to be harsh? (Remember that we are thinking here about what might have been happening tens of thousands of years ago as these mechanisms were evolving in ancestral humans.) And perhaps this weather forecast from the mother's body results in her baby being born with adaptations, such as a small body and a modified metabolism, which help it to cope with a shortage of food. This hypothetical set of adaptations has been called the 'thrifty phenotype'.[13] And perhaps these individuals with a thrifty phenotype, having small bodies and specialised metabolisms adapted to cope with meagre diets, run into problems if instead they find themselves growing up in an affluent industrialised society to which they are poorly adapted. That, at least, is the hypothesis.

People who grow up in impoverished conditions tend to have a smaller body size, a lower metabolic rate and a reduced level of behavioural activity.[14] These responses to early deprivation are generally regarded as pathological – just three of the many damaging consequences of poverty. But they could also be viewed as part of a package of characteristics that are appropriate to the conditions in which the individual grows up – in other words, adaptations to an environment that is chronically short of food, rather than merely the pathological by-products of a bad diet. Having a lower metabolic rate, reduced activity and a smaller body all help to reduce energy expenditure, which can be crucial when food is usually in short supply.

Now this conjecture might well be regarded as offensive. It could be seen as encouraging the rich to look complacently at their impoverished fellow humans, by arguing that all is for the best in this best of all possible worlds (as Voltaire's Doctor Pangloss would have had it). Merely to assert that every human develops the body size, physiology, biochemistry and behaviour that is best suited to their station in life would indeed be banal. The point, however, is not that the rich and the poor have the same quality of life. Rather, it is that, if environmental conditions are bad and likely to remain bad, individuals exhibit adaptive developmental responses to those conditions. To put it simply, they are designed to make the best of a bad job.

Of course, many of the long-term effects on health of a low birth weight may simply be by-products of the social and economic conditions that stunted growth in the first place. Ignorance and shortage of money make the prevention and treatment of disease more difficult; overcrowding, bad working conditions and poverty produce psychological stress and increase the risk of infection. People with little money have poorer diets, and adverse social or physical factors that foster depression and hopelessness increase the risks of disease.[15] In industrialised nations the poor and unemployed have more illnesses and die sooner than the affluent. But social and economic conditions do not account for everything, because the connections between low birth weight and subsequent health are still found among babies born in affluent homes.[11]

The thrifty phenotype hypothesis has led to a re-examination of what happens to people unlucky enough to have been born into an unusually harsh environment, such as a war. Towards the end of the Second World War, the German occupation forces in the Netherlands cut off the food supplies coming into the country. The population of much of the Netherlands suffered severe food shortages for six months. Babies born to mothers who suffered particularly badly from starvation during the final three months of their pregnancies were born with low birth weights. When these babies grew up, their capacity to deal with high levels of sugar was markedly reduced, as would be expected if they were adapted to a world containing little sugar.[16] One undesirable consequence for these individuals was an increased risk of developing diabetes, since they had actually grown up in a much richer, post-war environment. Another was that women who had been foetuses in the last third of pregnancy during the famine gave birth to children with lower birth weights than normal. This second-generation effect is important when considering the big differences in heart disease between adjacent countries.

The siege of Leningrad (now, once again, St Petersburg) by the German army in the Second World War lasted from the autumn of 1941 until the end of January 1944 and was one of the most gruelling sieges in history. The Germans, having invaded the Soviet Union in June 1941, approached Leningrad and almost completely encircled the city by late 1941, cutting off nearly all of its supply lines. The resulting starvation and disease, combined with German shelling, killed 650,000 of its inhabitants in 1942 alone. What happened to people who were born during the siege of Leningrad?

Fifty years later, scientists compared three groups of people: those whose mothers had suffered from malnutrition while they were pregnant during the siege; a second group born in Leningrad just before the siege began, who experienced starvation during their infancy; and a third group born at the same time but outside the siege area.[17] No differences were found in the incidence of coronary heart disease or diabetes between those whose mothers had been starved during pregnancy and those who were themselves starved in infancy. This result is still consistent with the thrifty phenotype hypothesis, since the nutritional state of people in Leningrad after the Second World War was generally poor compared with people living in the West. It could be argued that the mothers' nutritional weather forecasts, made to their unborn children during the siege, were on the whole reasonably accurate. Furthermore, although some people among all three groups grew into obese adults, those whose mothers had been pregnant during the siege had higher blood pressure than obese people in the other groups. This suggests again that they were less well adapted to a world markedly different from that of their foetal life – a world in which food was in rich supply.

The thrifty phenotype hypothesis suggests a novel explanation for another observation that has long puzzled nutritionists. People living in France have significantly less heart disease than people living in northern European countries. Genes alone cannot explain this difference in the incidence of heart disease because it persists in the face of widespread migration between countries. The difference is particularly noticeable when France is compared with Finland. The two countries have comparable average intakes of dietary cholesterol and saturated fat, yet the mortality from heart disease in Finland has for decades remained four to five times higher than that in France. People who grow up in the Perigord region of south-west France have a remarkably low incidence of heart disease, despite the fact that their diet is one of the fattiest in the world. What protects them from the ill-effects of the fat?

An analysis of dietary intake and coronary heart disease mortality statistics for forty countries has suggested that plant foods may provide some protection against the damaging effects of saturated fat and cholesterol in the diet.[18] Anther popular hypothesis is that wine (which the French famously drink in liberal quantities) protects against heart disease; red wine in particular has high concentrations of the anti-oxidant, flavonol, which is also found in plants such as tomatoes. The exclusive importance of this possibility has not yet been established,[19] and the thrifty phenotype hypothesis suggests another. It may be that a poor forecast during pregnancy prepares the developing offspring for a poor environment, while a good forecast prepares the individual to cope with a fatty diet later in life. The effect can carry over two generations and, as it happens, the French government supplemented the rations of pregnant mothers about fifty years before any other country introduced such measures.[20] With the improvement of nutritional conditions in other parts of Europe, the mothers' forecasts will increasingly correspond to reality and heart disease should drop to the low levels found in France.

If the thrifty phenotype hypothesis is true then the massive social experiment of the one-child policy in China, which we described earlier, could produce ill-effects, arising from a mismatch between body and environment. The only children resulting from the Chinese government's policy tend to be bigger, heavier and better nourished. But many of these children have been born to mothers who had low birth weights and thrifty bodies. If the thrifty phenotype idea is correct, the children may be at greater risk from the diseases of affluence. By the same logic, people who grew up in good environments may be at greater risk during periods of prolonged famine than those who were born as low birth weight babies. And perhaps children born to affluent parents are more likely to suffer adverse effects if they adopt rigorous diets in adulthood.

Growing People

Environmental conditions early in development have a significant impact on many other aspects of human biology, including size. People are getting bigger.[21] For decades now, the average height of men and women in industrialised countries has been steadily increasing. Although some of the height differences between people are due to genetic differences, the general trend for average height to increase is almost certainly due primarily to improvements in nutrition and, to a lesser extent, health. Hence, successive generations of the same family have grown taller despite having a similar genetic make-up.

The environmental improvements that have led to this general increase in height have affected males and females somewhat differently. Men have been growing taller faster than women. For example, the average height of men in Britain has been increasing at a rate of just over 1 cm every ten years, whereas the average height of women has been increasing at about one-third of that rate. In consequence, men are now relatively bigger than women than they were a century ago.[22]

Whilst the gap between the sexes has been widening, the average difference in height between social classes has remained roughly constant, with men from affluent professional homes being nearly 2 cm taller than men from manual backgrounds, and women from professional homes being 1·6 cm taller. In other countries the trends have been somewhat different. In Russia, where the improvements in nutrition have occurred more recently, the rate of increase in height lagged behind Britain but has been at almost three times the rates found in Britain in the last few decades. In the United States the trend towards ever taller offspring in successive generations, which started earlier than in Britain, has levelled off in recent years. These findings suggest an upper boundary on the effect of nutrition on human height.

Why should the average height of men be increasing faster

than that of women? It could be argued that in the past poor nutrition prevented people from achieving their full potential. Both men and women were prevented from growing beyond a minimum size and they were consequently closer together in height. Deprivation reduced both sexes to the same level. Improved environmental conditions, so this argument goes, have allowed a previously concealed sex difference in potential height to become apparent; if you were in the basement you could not get any lower, but above the basement the possibilities for differences between the sexes have expanded. This view is intuitively appealing, but it may not be the whole story. To see why, we need to consider what happens in other species.

Comparisons within and across other species of mammals reveal a connection between the relative body sizes of the two sexes and their reproductive behaviour. The larger the adult males are relative to the adult females, the less likely the males are to stay with one mate.[23] If the males of the species are much bigger than the females, they are unlikely to be monogamous. Why should this be the case? In a monogamous system of mating, both the father and the mother care for their offspring. In other mating systems, only one of the parents is primarily (or even wholly) responsible for parental care; in most mammals it is the female, but in some birds and many fish species the male cares for the offspring. In a polygynous mating system, one male mates with many females while other males suffer enforced celibacy. The reproductive success of individual males is therefore highly variable: a few males father many offspring while most males father few, if any. A mating system such as this, where the winner takes all, produces intense competition between males. One consequence is that, over the course of evolutionary history, males are selected to have larger bodies that are better equipped for competing and fighting with other males. Size and strength matter.

Across the animal kingdom, favourable ecological conditions

tend to be associated with polygyny, and harsh conditions with monogamy. This is because, with easy access to food, a single mother can rear her offspring without the help of the father. A harsh environment, on the other hand, will require the combined parental efforts of both father and mother if their offspring are to survive. Animals of one species may switch between polygyny and monogamy depending on the prevailing conditions. For example, the northern harrier, a predatory bird which feeds on voles, tends to be polygynous when voles are numerous and monogamous when voles are scarce.[24]

These associations between environmental conditions, body sizes and mating systems raise some provocative questions about humans. Is the type of marriage system found within a culture correlated with the difference in size between men and women in that culture? And is monogamy more common in societies living in harsh environments where food is in short supply? At first glance, the answer is disappointing, since the ratio of male to female height is roughly the same in both monogamous and polygynous human societies. However, a clear difference does emerge when monogamous societies are divided up into those in which the marriage system of monogamy is imposed by ecological conditions, and those found in more affluent conditions where monogamy is socially imposed, in the sense that it is a requirement of the culture. Men and women living in harsh environments where life is a struggle, such as the high Arctic and the edges of deserts, are more similar in size than those living in easier circumstances, and they are more likely to be monogamous.[25] This supports the adaptive hypothesis that, in poor conditions, a monogamous man caring for the offspring of one woman is more likely to have surviving children and grandchildren than a polygynous man who has children by many women.

The connection between environmental conditions, widely varied marriage systems and the extent of sex differences in height raises the intriguing thought that people might start life equipped for a range of alternative sexual lives. Maternal nutrition during pregnancy may trigger one of these alternative developmental pathways, just as reflected light affects the colour of the East African grasshopper or incubation temperature affects the sex of the turtle. The effect of such developmental predispositions on the subsequent historical growth of cultural practices such as marriage rules is unlikely to be inevitable or direct. Nevertheless, the existence of a predisposition could ease the general acceptance of a particular style of living. Once a mating system became common it could then be enshrined as a cultural rule, or even in law.

Pubertal Precocity

Besides the steady increase in adult height in developed nations, another striking trend in human development has been the fall in the age at which boys and girls reach puberty.[21] As adult heights have gone up, so the age at which people become sexually mature has come down. In Western countries the average age of menarche, when girls have their first period, has declined at a steady rate of eleven days per year over the last hundred years. All the other physical changes associated with puberty, such as the growth of breasts, the deepening of the voice in males and the appearance of pubic hair in both sexes, also appear earlier in development than they did a century ago.

In general, this downward trend in the age of puberty has started to level off throughout Europe, and in some cases has even shown signs of reversing.[26] As with the rise in adult height, the trend tended to occur in the higher socio-economic groups before poorer groups, and in industrialised countries before developing countries. Geographical variations are also found. The average age of puberty is slightly lower in southern Europe than it is in northern Europe, for example; thus, the average age of menarche in the early 1990s was 12·3 years in Spain, compared with 13·0 years in Denmark.[27]

As with the trend in adult height, the sexes differ. The average age of puberty (as marked by the sudden slowing of growth that occurs shortly after puberty) has declined more rapidly in girls than it has in boys.[21] This sex difference in the changing age of puberty has therefore worked in the opposite direction to that of height, where men have become relatively taller. Why has the age of puberty in girls apparently responded more rapidly to nutritional improvements than it has in boys? It may not be necessary to invoke an adaptive explanation. Girls develop sexually earlier than boys, so it could be that when nutritional conditions are bad, girls are simply forced to develop later and the gap between boys and girls is thereby reduced. Needless to say, another hypothesis can account for the sex difference in adaptive terms.

In a rich environment well supplied with food, females are able to start having children earlier in their lives. Females who responded to the nutritional conditions by starting to reproduce earlier would have more offspring than females who did not respond to those cues. Individuals who were equipped with the developmental flexibility to adjust the timing of their sexual maturation in response to environmental conditions would be at an advantage. This would set in train an evolutionary trend to establish a jukebox-like developmental response to the quality of the environment. The developmental rule would be: if conditions are good, become sexually mature early; but if conditions are poor, delay maturity. The developing male, however, must strike a balance between starting to breed as early as possible and developing the physical capacities to cope with competition from other males. A developing male who channelled all his resources into accelerating his sexual development, at the cost of remaining smaller and weaker, would be less well equipped to compete with other males. Without the physical size and maturity, a sexually precocious male would be unable to compete with adult males and would suffer in the process of trying. An optimum balance is struck when the young male has grown to a reasonably large size before he becomes sexually mature. This adaptive analysis fits the observation that good conditions have led to a larger acceleration in sexual development in females than in males.

Pursuing this logic suggests that even more sophisticated developmental rules may have evolved. Suppose an individual is born into a relatively good environment, but conditions suddenly deteriorate and the chances of having a long life consequently diminish. Under those circumstances, the individual would be better off going for a quick sprint to early reproduction rather than pacing itself for the long game.

Street Children

Darwinian hypotheses about human development responding, jukebox-like, to signals about the individual's future environment may seem speculative. Even so, young humans undoubtedly have a remarkable capacity to change their developmental tune in response to current conditions. When forced to do so for their own survival, even young children can quickly forgo their dependence on adults. Some children have little or no childhood at all. Around the world today, millions of children are being forced to grow up in conditions of abject poverty and without the benefits of what people in affluent countries regard as a normal childhood. They are forced to provide for themselves, and their relationship with adults is, at best, that of worker and employer rather than child and adult.

The 1989 United Nations Convention on the Rights of the Child, which has been ratified by more than a hundred countries, states that all children have rights to receive care, protection and a childhood. Article 31 specifically asserts every child's right to leisure and recreation. The convention is the most widely ratified human rights treaty in history, yet its stipulations are widely ignored. In many countries, including many signatories to the convention, a vast gulf exists between the noble aspirations expressed in the convention and the brutal reality of daily life for children. A British human rights group commented in a report that the convention was systematically and contemptuously violated by many countries, and that no countries violated it more energetically than those which had been quickest to sign.

At least one hundred million children under the age of eighteen – some as young as three – live and work on the streets in the developing countries of Asia, Africa and Latin America. According to estimates by UNICEF, 7·5 million homeless children live and work on the streets of Brazil alone. Although many street children retain some links with their families, they still spend most of their childhood on the streets, engaged in subsistence activities such as begging, shining shoes or selling trinkets to supplement their families' incomes. A third or more of street children around the world have no family and are homeless. They live in public parks, abandoned buildings or under bridges, often in the company of other street children. Their living conditions and the hazards they face in the day-to-day struggle for survival can be truly dreadful. A survey of street children in Guatemala, for example, found that every one of the children in the sample had been sexually abused at some point, mostly by relatives, and all of them regularly indulged in substance abuse – usually inhaling the vapour from glues and solvents. Such children often take to the streets to escape from violence and sexual abuse at home. They eke out a living under brutally harsh conditions, carving out a niche for themselves in the service economy, or resorting to crime or prostitution. Many have to support themselves, and sometimes younger children as well, when they are aged seven or younger. Their lives are often at risk, not only from disease but also from police and vigilante groups who regard them as a nuisance.

Despite the appalling conditions in which they grow up, these children show remarkable resilience. They are forced to behave like miniature adults. They work to earn a living, look after younger siblings in the absence of their parents, and co-operate or compete with other street children in their daily struggle for existence. They demonstrate remarkable adaptability in the face of terrible adversity and they certainly have a life that is about as different as it could be from the majority of children growing up in affluent societies. For those who survive, the long-term costs of a lost childhood are likely to be great, a point that we shall return to in Chapter 11.

Different Tunes

The notion of a developmental jukebox suggests that when the environment selects a particular tune, the outcome will be one of a few distinctive behavioural packages. Each of these tunes, or alternative lives, is adapted to the circumstances that triggered their development; so each alternative life will, according to this model, be distinct and appropriate to those particular conditions.

In considering the alternative ways in which any one person might develop, some common threads have started to emerge. The nutritional environment early in life has a long-term effect on body size and metabolism. It appears that the environmental cues for triggering one of these alternative lives are provided in the last few months of the mother's pregnancy. In the case of sexual development, the age of puberty is affected by the individual's nutritional state and environmental circumstances closer to the point at which a decisive commitment has to be made. Normally, children devote the long period between weaning and puberty to gathering the skills and knowledge they will need when adult. But the street children demonstrate that, when survival depends upon it, children can behave like adults, bypassing the normal slow course of childhood. In such ways the rivers of people's lives may flow down different channels.

7

Chance and Choice

> The whimsical effects of chance in producing stable results are common enough. Tangled strings variously twitched, soon get themselves into tight knots.
>
> Francis Galton, *Natural Inheritance* (1889)

Darwin's Nose

Chance may play a pivotal role in determining the course of a person's life. Charles Darwin recorded in his autobiography how his whole future hinged at one point on trivial and capricious circumstances:

> The voyage of the *Beagle* has been by far the most important event in my life, and has determined my whole career; yet it depended on so small a circumstance as my uncle offering to drive me thirty miles to Shrewsbury, which few uncles would have done, and on such trifles as the shape of my nose.[1]

Robert FitzRoy, the captain of HMS *Beagle*, the survey ship on which Charles Darwin spent five years, was not entirely joking when he considered rejecting Darwin because he thought the shape of Darwin's nose indicated a lack of determination. Like many others at the time, Captain FitzRoy believed passionately in phrenology. This was a school of pseudo-science, popular in the nineteenth century, which claimed to reveal a person's mental faculties and psychological traits from the contours of their skull. FitzRoy had similarly misplaced confidence in a related system that claimed to judge a person's character from the features of their face. Darwin's broad, placid looks seemed to betray a lack of inner resolution and insufficient energy. But despite his reservations about Darwin's nose, FitzRoy wanted above all a companion for the long voyage who was in his eyes a 'gentleman', a man from the same social class as himself. The first meeting between the two was a success; they took to each other and FitzRoy decided that, wrong nose or not, Darwin was a suitable person to join the voyage of the *Beagle*. From this expedition eventually stemmed Darwin's great book *The Origin of Species*.

Darwin was in later life a retiring man in poor health. When eventually he was persuaded to publish his theory of evolution by natural selection, he exploded a bombshell in the midst of Victorian society. It was as well for him that he had a formidable defender, Thomas Henry Huxley, who took on Darwin's religious and scientific adversaries with relish. Huxley had two famous grandsons, Julian and Aldous. Julian also became a biologist, while Aldous became a novelist and poet. Sixty-one years after the publication of *The Origin of Species*, Aldous Huxley pondered whimsically on the wildly improbable series of events that had given rise to him, or any other individual:

> A million million spermatozoa,
> All of them alive:
> Out of their cataclysm but one poor Noah
> Dare hope to survive.
> And among that billion minus one
> Might have chanced to be
> Shakespeare, another Newton, a new Donne –
> But the One was Me.

Huxley's biology was slightly off-target, since the chances are that none of his father's spermatozoa would have had whatever cocktail of genes was needed to produce a Shakespeare, a Newton or a Donne. It is also highly unlikely that any of his father's possible children would have grown up in exactly the right sort of environment to produce the great men in question, even if the right genes had been in place. Nevertheless, the point is clear. Chance plays a major role in shaping each person's development, throughout their lifespan.

The sperm cell that fertilised the egg that became Aldous Huxley was equally likely to have carried an X chromosome as a Y chromosome; so Huxley had the same chance of becoming a girl as a boy. He might also have had Down's syndrome, his mother might have died at his birth or he might have been born early. Premature birth, which can happen for a variety of often unpredictable reasons, brings with it a set of problems that may have long-term consequences. Premature babies have a limited ability to feed by mouth, reduced gastric capacity and limited digestive capabilities. They start life at a disadvantage.[2]

The luck of the draw continues with the circumstances into which an individual is born. Socio-economic status has a major bearing on all aspects of life, including health and life expectancy. At all ages, individuals from lower socio-economic groups have substantially poorer health and higher mortality rates than individuals in the same society from wealthier backgrounds. In particular, they have more heart disease, more strokes and more cancer.[3] The higher incidence of illness stems partly from differences in behaviour: lack of breastfeeding, smoking, lack of physical exercise, obesity and poor diet are all more prevalent among the poor. But whatever the causes of the diseases of poverty, they are mostly beyond the control of the individuals concerned.

Other people obviously have important influences on the

course of any one individual's life, but just who will be encountered during a lifetime is again a matter of chance. Like Darwin, Aldous Huxley and many others, Simone de Beauvoir mused on the chance events that combined to make her the person she was. In particular, she pondered on the stroke of fortune that brought her together with the philosopher Jean-Paul Sartre when they were students at the Sorbonne in the late 1920s. She regarded that meeting as the most important event in her whole life. Sartre was to be her lifelong companion and a crucial influence on her personal and intellectual development, as de Beauvoir recorded in her autobiography:

> How is a life formed? How much of it is made up by circumstances, how much by necessity, how much by chance, and how much by the subject's own options and his personal initiatives? . . . a thousand different futures might have stemmed from every single movement of my past: I might have fallen ill and broken off my studies; I might not have met Sartre; anything at all might have happened. Tossed into the world, I have been subjected to its laws and its contingencies, ruled by wills other than my own, by circumstances and by history: it is therefore reasonable for me to feel that I am myself contingent . . . Chance, in one form or another, helped to fill my life with people . . . chance favoured me extraordinarily in placing Sartre upon my path . . . How should I have developed if I had not met Sartre? . . . I cannot tell. The fact is that I did meet him and that was the most important event in my life.

Storm and Stress

Events over which people have no control can have devastating effects on the course of their lives. Extreme and prolonged shortages of food have been a recurring feature of human history and continue to the present day. The earliest recorded famines occurred in the Middle East nearly six thousand years ago. Rome was hit by famine in 436 BC, when thousands of starving people committed suicide by throwing themselves into the River Tiber. Britain suffered numerous famines in the Middle Ages; in the year 1235, for example, some 20,000 Londoners died from starvation. India has experienced famine throughout its recorded history. One of the worst occurred in 1769–70, when ten million people out of a total population of thirty million died from starvation and diseases such as malaria, smallpox and cholera, to which the starving people were more vulnerable. The disastrous Irish famine of 1846–51, triggered by a fungal disease in the main food crop, the potato, forced two million people to emigrate. Famine continued through the twentieth century. More than three million people starved to death in China during a famine in 1928–9. In recent years, millions have died in famines in Africa, brought on by drought and civil war. These tragedies do not just cause death. They also leave a prolonged mark on the living.

Someone who experiences terrible events such as war or famine may be left psychologically scarred. In the First World War a condition known as 'shell shock' was recognised. Some of the sufferers were shot for cowardice. But the condition also provided a powerful stimulus to the then new subject of psychiatry. Men who had been in combat became hypersensitive to sudden noises and movements. They had difficulty in getting to sleep and, when they did eventually sleep, had appalling nightmares. During the day, their jumpiness and irritability could turn into physical violence. These men were often treated as insane and placed in institutions. Their illness was, at least in the early days, attributed to the physical effects of shells exploding around them, rather than to psychological damage.

The same syndrome recurred in soldiers fighting in the Second World War, when it was known as 'battle fatigue'. The usual treatment was rest, food and sedation. Sufferers benefited from staying with their combat units, where the comradeship provided some security. Nevertheless, long-term studies of veterans consistently found that the ill-effects of combat stress could persist in some individuals for fifty years or more.[4] A large number of veterans of the Vietnam War suffered long-term behavioural and drug-abuse problems.[5]

Worldwide, around two million households annually experience damage or injuries from fires, floods, earthquakes, volcanic eruptions, hurricanes, tornadoes and other natural disasters.[6] To these must be added the huge number of civilians caught up in wars, bombings and other politically motivated disturbances, and the survivors of violent crime and transport disasters. Many of the survivors of such traumas, who may be bereaved and witnesses to horrible suffering, behave in similar ways to soldiers suffering from battle fatigue. The syndrome is now known as post-traumatic stress disorder (PTSD).[7] The effects of the traumas dissipate over time in most people, but in some individuals, for a variety of reasons, they may last for years. The main symptoms of PTSD are heightened sensitivity to anything that might be stressful, coupled with a feeling of numbness towards the world. The sufferers have intrusive memories and sleep disturbances, with recurring nightmares about the event that traumatised them. Probably as a result of the chronic stress, they are more likely to be afflicted by disease.

Why do some individuals suffer such persistent effects while others recover relatively quickly? Not surprisingly, people who experience the most intense trauma are more at risk, other things being equal, of suffering long-term psychological damage. Among children caught up in the prolonged Lebanese civil war, those who had experienced the most severe trauma exhibited the clearest and longest-lasting signs of PTSD.[8] Similarly, a study of former political prisoners in Turkey found that those who had been tortured most severely were most affected in the long run.[9] Gauging the severity of the trauma at the time it happens is difficult, but objective measures are sometimes possible. Among a large group of children who lived through an earthquake disaster in Armenia in 1988, those who were closest to the epicentre of the earthquake showed the greatest signs of post-traumatic stress disorder eighteen months later.[10] Often the worst effects are produced by bereavement.

These cases are noteworthy because the traumas changed individuals' lives so profoundly. The traumas were not sought and were usually unexpected. They may exert some of the same prolonged effects on behaviour as early experience, a point to which we shall return in later chapters. Sometimes, doubtless, the chance events that seem so important to those who experience them are imbued with too much significance; in reality they may have done little to alter the overall course of a life. Maybe, even without his experience of the *Beagle* voyage, Darwin would nevertheless have brought his remarkable mind to bear on the matters that preoccupied him throughout his life, arriving at the same outcome by a different route. Maybe Simone de Beauvoir would have been the person she was even without meeting Sartre. But this does seem unlikely.

Chance is not always what it seems. It is hard to dispute that some accidents have nothing to do with the actions or personality of the person who suffers their ill-effects. But sometimes people inadvertently invite misfortune upon themselves. Accidental injuries may be more likely to happen to some people rather than to others. Those with neurotic personalities are more likely than others to have unpleasant experiences. They are more inclined to self-destructive behaviour such as over-eating, smoking excessively, drinking large quantities of alcohol, using other addictive drugs or committing suicide.[11] The non-random aspect of 'chance' was described by Samuel Butler in *The Way of All Flesh*:

Fortune, we are told, is a blind and fickle foster-mother, who showers her gifts at random upon her nurslings. But we do her a grave injustice if we believe such an accusation. Trace a man's career from his cradle to his grave and mark how Fortune has treated him. You will find that when he is once dead she can for the most part be vindicated from the charge of any but very superficial fickleness. Her blindness is the merest fable; she can espy her favourites long before they are born.

What's in a Name?

Parental divorce, childhood illness, physical appearance, particular teachers, natural disasters, involvement in war – these and countless other unpredictable influences may all play their part in shaping an individual's life. Even a name can matter.

Shakespeare famously mused on the inconsequential nature of names ('What's in a name? that which we call a rose/By any other name would smell as sweet.') The properties of a rose are obviously unaffected by the arbitrary name that humans give to it. But the same may not be true of a person. The links that are occasionally observed between someone's name and their trade or profession have been a source of amusement throughout the centuries. In the medical profession Lord Brain was a neurologist and two experts on incontinence were called Splatt and Weedon. Dr Smiley was an orthodontist. Drs Heavens, Starr and Stella Law were astronomers, Dr Mountain a geologist and Dr Fish a marine biologist. The daughter of the head gardener of King's College, Cambridge married a man called Greenhouse. And so we could go on. Are such cases mere coincidences – the statistically inevitable connections in an ocean of randomness – or might someone's name actually have an influence on their choice of career? The *New Scientist*, tongue-in-cheek, described such cases as 'nominative determinism'. With mere anecdotes it is easy to construct a story out of nothing, simply selecting the cases that fit the argument and ignoring those that do not. More substantial evidence would come from analyses that guard against bias by introducing a comparison. For example, the effect would be established if family names such as Sparks were more common among people working in the electricity industry than in the water industry, while people with names such as Leak were more likely to be plumbers than electricians.

First names too may have some connection with an individual's character. A study at Harvard University found that students with unusual first names were more likely to drop out from their courses than those with common names. This might have been because the choice of an unusual first name reflected a parental temperament which in turn influenced the oddly named children's subsequent behaviour. In some cultures, however, a child's first name is not always a matter of parental choice. In the West African Ashanti tribe, boys are named after the day of the week on which they were born, and the names are associated with particular personality characteristics. *Kwadwo* (Monday) is supposed to be quiet, retiring and peaceful, whereas *Kwaku* (Wednesday) is quick-tempered, aggressive and a trouble-maker. A study found that the personalities associated with the names had some basis in reality. *Kwadwo* boys, carrying the peaceful name, were significantly less likely to appear as delinquents in juvenile courts, whereas *Kwaku* boys, associated by their names with trouble, were significantly more likely to be involved in offences against the person.[12] This looks like a real case of nominative determinism, where individuals come to act in a way that is suggested by their names. People do not usually choose their names, but their decisions which set them off in one particular direction may be subtly influenced by the chance that they happened to be labelled in a particular way.

Choice

At many moments in the course of their lives, individuals make decisions for themselves or have them made by others. These decisions have profound ramifications on the subsequent course of their lives. During the industrialisation of many countries, millions of people decided to leave the places where they had been brought up and move to the towns. In Russia the main migration came at the end of the nineteenth century. In his book about the Russian Revolution, *A People's Tragedy*, Orlando Figes describes how a young peasant found the poverty and monotony of village life unbearable. Virtually any employment in the city seemed exciting and desirable compared with the hardships of peasant life. Becoming a clerk or a shop assistant was seen by the younger peasants as a move up in the world. For the young women who found themselves at the bottom of the patriarchal pile, working as a domestic servant in the city offered them a much more attractive and independent life. Whatever the reasons for choosing the move from the impoverished conservative world of icons and cockroaches to the urban world, standing for progress, enlightenment and liberation, the experience of the city transformed the thinking of most Russians peasants who made the change. And so they repeated a pattern that had already occurred in many parts of Europe and would occur in other parts of the world throughout the twentieth century.

Parents in some societies choose their child's spouse, devoting much thought, effort and negotiation to the process. Some parental decisions trap the child into an unenviable life. Parents might simply follow tradition, as in the old Chinese practice of binding women's feet so that they were unable to walk with ease. Parents might have their daughter's clitoris and labia removed, greatly reducing her capacity for sexual pleasure. Female circumcision is still widely practised in parts of the Middle East, Africa, Asia and South America. The ritual is viewed as part of a religious or ethnic tradition and as a necessary rite of passage into adulthood. Some parents, through poverty, sell their children into slavery or prostitution. A Jehovah's Witness might refuse a life-saving blood transfusion. Conscious or unconscious choices can shape an individual's life, but they are not always adaptive.

Fiction is rich with examples of people who, by their own conscious efforts, decide to better themselves, following the tradition that men made money and women married where money was to be found. John Braine's *Room at the Top*, one of the British 'Angry Young Man' novels of the 1950s, relates the rise and rise of Joe Lampton, a ruthlessly ambitious and opportunistic young man who decides to change the course of his life. Through sheer will and cynical determination, Lampton prises himself out of the genteel poverty of life as a young local government employee in post-war northern England and instead inserts himself into his chosen life of wealth and glamour. This radical change reflects a deliberate decision that Lampton takes in a moment of insight. He is sitting in a café, contemplating his dull prospects as a newly employed bureaucrat, when he catches sight of an unknown young man of obvious wealth who is driving an Aston-Martin sports car and has a beautiful woman by his side. The experience changes the course of Lampton's life and sets in train a chain of events that lead to the attainment of his goals, albeit at considerable personal cost to himself and those around him. He goes on to seduce and marry the daughter of a wealthy businessman, casting aside his unhappily married lover. Lampton later recalls the crucial moment in his personal development as he watched the Aston-Martin driver through the café window:

> For a moment I hated him. I saw myself, compared with him, as the Town Hall clerk, the subordinate pen-pusher, half-way to being a zombie, and I tasted the sourness of envy . . .
>
> As I watched the tail-end of the Aston-Martin with its shiny new G.B. plate go out of sight I remembered the second-hand Austin Seven which the Efficient Zombie, Dufton's Chief Treasurer, had just treated himself to. That was the most the

> local government had to offer me; it wasn't enough. I made my choice then and there: I was going to enjoy all the luxuries which that young man enjoyed. I was going to collect that legacy . . . I was moving into the attack, and no one had better try to stop me. General Joe Lampton, you might say, had opened hostilities.

Charles Darwin agonised about whether or not to marry – a matter over which he presumed he had some control. He drew up a list of arguments for and against marriage. If he did not marry he would be able to read in the evenings, he wouldn't be forced to visit and receive relations (such a terrible loss of time, he felt) and marriage might mean banishment to the country. But if he did marry, he would receive care, comfort and company and he would have the pleasure of children. But then children would be such a source of anxiety . . . Having considered the bleak prospect of a solitary old age, Darwin urged himself to trust to chance and to keep a sharp lookout. After committing these thoughts to paper in 1838, he started to court his cousin Emma. They later married and lived happily together for over forty years.

Sometimes the choices that people make seemingly stem from playing out in their minds the possibilities that are available to them. They may start to do so by putting themselves in others' shoes. The point is illustrated in a list of entertaining challenges to the imagination:

> What is it like to work at McDonald's? To be thirty-eight? . . .
> What is it like to climb Mount Everest? To be an Olympic gold-medal winner in gymnastics? . . .
> What would it be like to be a good musician? To be able to improvise fugues at the keyboard? To be J.S. Bach? . . .
> What is it like to believe the earth is flat? . . .

> What is it like to hear English (or one's native language) without understanding it?
> What is it like to be the opposite sex? . . .[13]

Empathy is in an important part of the way that humans are able to deal successfully with each other (and, increasingly, to give the benefit of the doubt to other animals). The faculty of empathy also enables humans to imagine other lives for themselves; having done so, they may then seek to change themselves. Some transsexuals first dream about what their lives would have been like if they had been born as a member of the other sex and then, with the aid of surgery and hormonal treatment, seek to experience their dream. The British writer Jan Morris was born a man, the youngest of three brothers. As the young journalist James Morris he accompanied the 1953 British expedition to climb Mount Everest and was the first to break the news that it had been conquered. He married and had children, but since early childhood he had felt himself to be a woman trapped in a man's body: 'I was perhaps three or four years old when I realized that I had been born into the wrong body, and should really be a girl. I remember the moment well, and it is the earliest memory of my life.'

After living the life of a male transsexual until middle age, James emerged from a surgeon's clinic in Casablanca as Jan. Her adoption of the female gender is memorably recounted in her autobiography *Conundrum*. Jan Morris wrote:

> Nobody really knows why some children, boys and girls, discover in themselves the inexpungable belief that, despite all the physical evidence, they are really of the opposite sex . . . That my inchoate yearnings, born from wind and sunshine, music and imagination – that my conundrum might simply be a matter of penis or vagina, testicle or womb, seems to me still

> a contradiction in terms, for it concerned not my apparatus, but my *self* . . . it all seemed plain enough to me. I was born with the wrong body, being feminine by gender but male by sex, and I could achieve completeness only when the one was adjusted to the other . . . So I do not mind my continuing ambiguity. I have lived the life of a man, I live now the life of a woman, and one day perhaps I shall transcend both – if not in person, then perhaps in art, if not here, then somewhere else.

Thomas Hardy's Tess, in *Tess of the d'Urbervilles*, makes the following assertion of her uniqueness and individuality, rebelling against the assumption that the course of her life was predetermined at birth by her lineage and circumstances:

> What's the use of learning that I am one of a long row only – finding out that there is set down in some old book somebody just like me, and to know that I shall only act her part; making me sad, that's all. The best is not to remember that your nature and your past doings have been just like thousands' and thousands', and that your coming life and doings'll be like thousands' and thousands'.

Scholars once debated Thomas Hardy's determinism. Some argued that his characters were mere puppets in the hands of malign fate (such as Tess or Jude the Obscure). Others argued that, on the contrary, Hardy was scrupulous in giving his characters choice; the apparent determinism arose because the characters make the 'wrong' choices. The argument does not end here, of course, because an explanation is still required for choice itself. Why prefer one route rather than another? This brings us back to the subtle interplay between biology and conscious choice.

How Much Free Will?

When somebody makes a conscious choice, do they *really* know what they are doing? The presumption of law, morality and, indeed, common sense would be that they do. But even legal systems, not famous for recognising grey areas, accept pleas of diminished responsibility in criminal trials. The deterministic character of modern biology seems to go further and subvert the common-sense view that individuals can make free choices. An obvious retort is that a well-designed brain should respond to the consequences of behaviour; if an understanding of the likely consequences can be achieved without actually performing the act, then a person who knows that they will be rewarded or punished for certain acts is bound to be influenced by that knowledge. A brain designed in that way invites the evolution of societies with explicit social approval of certain activities and explicit disapproval of others. The rules for what is or is not acceptable may be arbitrary, but only fools and the brave will ignore them. The point is that people do make sensible, well-planned decisions. A proper understanding of biology embraces free will.

Nevertheless, grey areas do remain. The choices and decisions that people make during their lives need not involve conscious thought. People may choose a course of action without knowing why or reflecting on what they do. It just seemed right at the time. Those who wish others to buy their wares or use their services seek to manipulate choice. Supermarkets tempt customers to buy things they did not know they wanted by skilful arrangements of displays or by wafting the aroma of freshly baked bread through the ventilation system. As people become familiar with certain commercial brands through repeated exposure to advertisements, they will prefer to buy them – mainly because they *are* familiar. For instance, foreign words made familiar to students by printing them on T-shirts were preferred by the students over foreign words they had not seen before. Similar effects are obtained in controlled laboratory conditions using Chinese ideographs.[14]

People with particular sorts of damage to the pre-frontal cortex of the brain provide living proof that rational thought and conscious analysis are insufficient to function well in the real world. Their ability to make even apparently straightforward decisions – especially decisions involving social or personal subtleties – is severely impaired, despite their normal performance in almost any test of intelligence, memory or reasoning. They remain engaged in a trivial activity, neglecting the important activities of daily life. The neurologist Antonio Damasio described eloquently in his book *Descartes' Error* how these patients seem to lack the biasing that the emotions normally provide when making decisions about budgeting time. Elliot was one such patient who had a tumour removed from the critical brain area. He returned to work after recovering from the surgery, but would not stick to any schedule set for him. When sorting documents he would spend a whole day reading one letter. He lost his job and by degrees his whole life fell apart.[15]

Damasio argued that the emotions are crucial in everyday life. He and his colleagues also showed the value of hunches, in a gambling task designed to simulate real-life decision-making. Volunteers could win or lose facsimile money by choosing cards from various decks. Unknown to them, choosing cards from 'bad' decks led to overall losses, while 'good' decks produced overall gains. After they had experienced some losses, normal people began to choose cards from the 'good' decks before they had consciously realised which strategy worked best. After further experience they became consciously aware of the difference between the good and the bad decks of cards and ceased to play on the basis of hunch.[16] This study indicated how non-conscious biases advantageously guide behaviour before conscious knowledge or rational analysis comes into play.

Inchoate preferences guide and facilitate conscious evaluation and reasoning.

Tangled Threads

Chance and choice combine to shape people's lives in ways that are often impossible to predict. The possible tangles are limitless. Chance and choice are also intertwined. It was by chance that Simone de Beauvoir met Sartre, but it was by choice that they lived together. Buying a lottery ticket involves choice (except for the most addicted gambler) but winning is pure chance. Suddenly having a large sum of money may change a person's life in ways that they had not foreseen.[17]

The experiences that shape a life may be gained in many ways, through chance or by choice. In the developmental kitchen many things matter in making a dish. Sometimes omitting an ingredient from a recipe or taking a shortcut in the preparation has no obvious effect; but in other cases even a small deviation from normal practice can have consequences that are disastrous or sublime. The seemingly chaotic effects are hard to predict, sometimes closing down a person's willingness to experience change, as in the case of people with PTSD, and sometimes liberating them in ways they would not have expected. The impact of chance and the opportunity for choice may, nevertheless, be channelled by an important aspect of human biology and these influences may be especially likely to occur at particular stages in development. This is the topic to which we turn in Chapter 8.

8

Sensitive Periods

> There is always one moment in childhood when the door opens and lets the future in.
>
> Graham Greene, *The Power and the Glory* (1940)

Vulnerable Times

In 1961 the world was shocked to learn that a widely used wonder drug had horrible side effects. The drug, called thalidomide, had been marketed in forty-six countries, helping thousands of pregnant women. It was taken as a sedative in the first three months of pregnancy and relieved morning sickness; at this early stage in the unborn baby's development the limbs start to sprout from buds. The drug sometimes had a pronounced effect on these points of growth. Babies were born with severe limb deformities: some had no arms, some had no legs with toes growing from their hips, some had limbless trunks. While a few children had intellectual disabilities, most were normal in their behaviour and subsequently showed triumphantly how people can cope with terrible adversity. The drug did not affect their mothers' bodies, and thalidomide became one of the most notorious of the poisons that only cause damage at a certain stage in human development.

The physical development of organs is a tight sequence of

events. Each organ has a period of maximum growth, during which it is particularly vulnerable to any disruption. In humans much organ formation occurs during the first eight weeks after conception. It is therefore particularly important for pregnant women to avoid exposure to toxic chemicals during this period. The developing foetus is also vulnerable to damage by radiation and diseases such as German measles. The harmful effects of even modest amounts of radiation were revealed in Japanese women who were pregnant when atomic bombs were dropped on Hiroshima and Nagasaki in 1945. Some of these women gave birth to babies with deformities such as cleft palates, stunted limbs and underdeveloped brains. Similar effects were found in a small proportion of women who had been exposed to medical X-rays during pregnancy in the days before the harmful effects of radiation were fully appreciated.

Excessive consumption of commonly used drugs such as tobacco and alcohol during pregnancy can also damage the foetus. The babies of some alcoholics develop a distinct disorder known as foetal alcohol syndrome (FAS), the signs of which include low birth weight and later growth retardation, small head circumference, mental retardation and, in many cases, congenital heart disease. In less extreme cases, the babies of some alcoholics display slow development, learning disabilities and hyperactivity.[1] FAS is found in around three babies in every thousand born alive in the United States.[2]

In the 1920s Charles Stockard, a leading embryologist of his time, observed that embryos of many species could resist oxygen deprivation at some stages of their development, but that at times of rapid growth oxygen deprivation could turn them into monsters, sometimes with two heads. He called these stages of increased vulnerability 'critical moments'. This term, or variations on the same theme, has been used in most scientific accounts of the long-lasting effects of events that occur early in life. We shall use the term 'sensitive period' because, in most cases, such periods of increased susceptibility begin and end gradually rather than abruptly and depend on local conditions.

The pathological processes set in train by thalidomide, X-rays, alcohol or oxygen deprivation during early development are wholly unnatural and obviously have no biological use for the individual. Do they reveal anything about the mechanisms involved in building a normal body? The answer is that usually these examples point to vulnerability at times of rapid growth. Nevertheless, they may reveal those periods in development when the organism requires information from its environment in order to develop normally. This has some relevance to behavioural and psychological development. As we discussed in Chapter 6, forecasts about the state of the environment, provided by the mother's state during the later stages of pregnancy, are likely to play a role in normal development, determining which of several alternative developmental courses her child will adopt.

The developing foetus may be affected if the mother is severely stressed, but in this case it is less obvious whether her state provides useful information. Much of the evidence on how developing offspring are affected by maternal stress comes from animal studies. In rats, pre-natal stress in pregnant females results in greater emotional reactivity, anxiety and depressive-like behaviour in their offspring.[3] Similar effects have been found in primates.[4] In one study, rhesus monkeys which had been born to stressed mothers were subjected to mildly challenging conditions at four years of age. Their behavioural responses differed significantly from those of monkeys born to unstressed mothers. In particular, they were more alarmed by separation from cage mates, more clinging, less exploratory and less playful.

The building of a body is about co-ordination and timing – providing particular cues at the right moments. The male hormone testosterone serves such a role early in life, sending out signals to the growing body to start building male structures like a penis, and triggering subtle changes in brain development that will eventually be required for male behaviour. Later in development, testosterone plays a different role in co-ordinating the various aspects of sexual behaviour in the adult male. In rare cases, a woman may produce substantial quantities of testosterone in her adrenal glands. If a woman suffering from this disorder is pregnant with a daughter, the testosterone may pass through the placenta, reach her unborn child and induce in her the development of male characteristics.[5] In such cases the abnormalities cast light on what usually happens when sexual characteristics develop normally. Like comparisons between species, comparisons between the normal and the abnormal draw attention to aspects of normality that would otherwise seem unremarkable.

Building the immune system, which protects the body against viruses, bacteria and bigger parasites, is another extraordinary feat of biological orchestration. The immune system must be immensely versatile in recognising new threats. It must also be able to distinguish unerringly between the cell surfaces of foreign invaders and those of the body it serves – in other words, between self and non-self. The more accurately the immune system performs both these functions, the better able is the individual to avoid infection and disease. If the immune system is too lax or haphazard, the body will be overrun by pathogens or parasites that cause disease and manipulate the host's body to serve their own reproductive ends. If, on the other hand, the immune system is too vigorous and indiscriminate, it may attack tissues in its own body, causing auto-immune diseases such as rheumatoid arthritis.

The capacity of the immune system to recognise foreign invaders and respond appropriately is enhanced by early experience. The human immune system evolved in an environment in which children were predictably exposed to a wide range of pathogens in early life and, presumably, were not stopped from putting things into their mouths. In affluent families today, it seems that anxious parents protect their children too much. Children's susceptibility to some diseases in later life could result from their being deprived of what would have been normal experience in the environment in which humans originally evolved.[6]

Similarly, breastfeeding enhances the baby's protection against infection.[7] For example, a Scottish study tracked a large sample of mothers and infants from birth to two years of age.[8] Babies who had been breast-fed for the first three months or more of life had significantly less gastrointestinal illness, fewer hospital admissions and lower rates of respiratory illness than babies who had been bottle-fed from birth. The reduced incidence of gastrointestinal illness persisted after the end of breastfeeding. Babies who were breastfed for less than three months had similar rates of gastrointestinal illness to bottle-fed babies. When the same children were examined between the ages of six and ten, those who had been breastfed were found to have a markedly lower risk of respiratory illness. Breastfeeding is also associated with higher levels of cognitive and visual abilities.[9] This may arise in because breast milk contains nutrients (docosahexaenoic and arachidonic acids) that are essential for brain development but are usually not abundant in formula milk. The point about the benefits of breastfeeding is they have to be experienced early in life. Obviously these benefits are not easily obtained later in development and, even if they were made available, would arrive too late to do much good because the window of opportunity had closed.

Something Nasty in the Woodshed

The crucial importance of early experience in the development of each individual's personality, values and character is deeply embedded in fiction and folk psychology. The central role of early experience features in many educational theories of different vintages. As one of the writers of the Old Testament asserted: 'Train up a child in the way he should go: and when he is old, he will not depart from it.' A similar maxim was famously attributed to the Jesuits: 'Give me a child for the first seven years, and you may do what you like with him afterwards.' Ideas about the importance of early experience in shaping people's lives became so fashionable in the first half of the twentieth century that they were mocked. Here is Stella Gibbons's famous tease in her comic classic *Cold Comfort Farm*:

> Aunt Ada Doom sat in her room upstairs . . . alone.
>
> There was something almost symbolic in her solitude. She was the core, the matrix, the focusing point of the house . . . and she was, like all cores, utterly alone . . .
>
> When you were very small – so small that the lightest puff of breeze blew your little crinoline skirt over your head – you had seen something nasty in the woodshed.
>
> You'd never forgotten it.
>
> You'd never spoken of it to Mamma – (you could smell, even to this day, the fresh betel-nut with which her shoes were always cleaned) – but you'd remembered all your life.
>
> That's what made you . . . different.
>
> That – what you had seen in the tool-shed – had made your marriage a prolonged nightmare to you.
>
> That was why you had brought your children into the world with loathing. Even now, when you were seventy nine, you could never see a bicycle go past your bedroom window without a sick plunge at the apex of your stomach . . . in the bicycle shed you'd seen it, something nasty, when you were very small.
>
> That was why you had stayed here in this room. You had been here for twenty years . . .
>
> Outside in the world there were potting-sheds where nasty things could happen.
>
> But nothing could happen here . . .

You told them you were mad. You had been mad since you saw something nasty in the woodshed, years and years ago . . .

The woodshed incident had twisted something in your child-brain, seventy years ago.

And seeing that it was because of that incident that you sat here ruling the roost and having five meals a day brought up to you regularly as clockwork, it hadn't been such a bad break for you, that day you saw something nasty in the woodshed.

Many aspects of human behaviour are easily changed by learning in adulthood, so it might seem implausible that early experience should have an especially dominant effect. Some types of learning involving rewards and punishments readily affect how any adult behaves. Yet not all forms of learning occur with equal facility at all ages.

Learning to Speak

A familiar example of a special time for learning is the development of language. The human ability to acquire language is truly remarkable and too easily taken for granted. Children who have been massively deprived and then rescued show remarkable resilience, and yet the difficulty of learning a language at a later age than normal often proves one of their greatest stumbling blocks. In one typical case, two Japanese children (a brother and sister) were taken into care after they had been found living in conditions of extreme social isolation and deprivation. When they were discovered, the children's chronological ages were five and six years, but their psychological development was characteristic of infants less than one year old. As adults they appeared normal in some respects, but their language was impaired, as were their social and cognitive abilities. Sorting out cause and effect here is impossible, but their lack of exposure to normal language early in life probably made normal linguistic development impossible for them.[10]

Many other strands of evidence point to the crucial importance of acquiring experience of language early in life. An accent derived from childhood experience is particularly difficult to lose in a newly acquired second language, and some vowel and consonant sounds in the new language may not be recognised at all. A well-known example is the confusion by native speakers of Japanese and Chinese of the 'r' and 'l' sounds in spoken English. Conversely, people who speak English as a first language almost invariably fail to detect a consonant used in the Thai language, which lies in the phonetic slide between 'd' and 't'. Even within the European languages, English speakers from the United States cannot distinguish between various forms of 'a' used in Swedish. The Swedes, needless to say, have no problems with their own language.[11]

Babies start life with the ability to recognise and distinguish between the full range of different vowel and consonant sounds used in all human languages. The physical boundaries between those sounds are blurred, but the human ear perceives them as sharply defined categories. Exposure to a particular language during a sensitive period early in life alters the way in which babies categorise sounds. At 4–6 months of age infants can detect subtle phonetic differences between syllables in unfamiliar languages as well as their native language. But by 10–12 months of age their discrimination has narrowed as a result of experience, so that they can only tell the difference between sounds that are commonly used in their own native language.[12] Mothers speak in a way that assists this narrowing process. When talking to their infants they produce more extreme versions of the vowel sounds than when talking to adults. This helps to expose their child to the linguistic building blocks of their native language.

Children raised from birth to speak two languages can do so

with ease. But learning a second language in adolescence or adulthood is much more difficult, and adults rarely become completely fluent in a second language. In one study, native Chinese speakers who had moved to the United States and learned English as a second language were tested on their ability to understand the way questions are formulated in English.[13] The older the individuals were on arrival in the United States, the worse they performed as adults. Similarly, native Russian speakers have difficulty using the definite article in English speech, even after being immersed in the English language for many years. The great ballet dancer Mikhail Baryshnikov defected to the West in his twenties; thirty years later he described his childhood experiences in a form of English that is characteristic of many native Russian speakers:

> In Russia, dancing is part of happiness in groups. Groups at parties, people dancing in circle, and they push child to center, to dance. Child soon works up little routine . . . and soon make up some special steps, and learn to save them for end, to make big finale. This way, child gets attention from adults.[14]

Most experts agree that speech perception and accent develop far more readily during childhood than they do later in life. Adults trying to acquire a second language easily seem to have missed the train.[15] Even so, considerable argument still revolves around whether early exposure is required for *all* aspects of language acquisition. Extensive vocabularies can be learned as readily, or more readily, in adult life than as a child.

In normal circumstances, all individuals acquire language perception, sound production and sentence construction early in life. Even so, it might seem to make biological sense to retain the full ability to learn language throughout life. So why do some aspects of this ability fall away after childhood? One plausible answer is that, in terms of the energy that it requires, brain tissue is expensive to run. If our ancestors received little or no benefit from retaining the full capacity to learn language into adulthood, individuals would be at a disadvantage if they retained the neural scaffolding required to build the one and only language they would ever need. They would have to provide energy for a part of the brain that no longer had a useful function.

The concept of neural scaffolding used for building human language is supported by evidence about how the brain is used in language. When a part of the frontal lobes known as Broca's area is damaged by stroke, for example, the person loses his or her capacity to speak. A form of brain scanning, known as functional magnetic resonance imaging (MRI), has revealed that the brain area that becomes active when people silently 'speak' in their native tongue is different from the area that becomes active when they silently 'speak' in a language they have acquired later in life. As the MRI scanning took place, volunteers were asked to imagine themselves speaking about events that had occurred the previous day, either in their native language or in their second language. (Silent 'speech' was used in order to keep to a minimum any head movements that would have made the brain scans imprecise.) In people who had learned their second language early in life, the same part of the brain was active when both the native and second languages were 'spoken'. However, in people who had acquired the second language in adulthood, a different part of the frontal lobe, almost a centimetre away from the native language area, became active instead.[16]

These remarkable findings suggest that acquiring one or more languages in childhood involves one set of brain structures, while acquiring a new language later in life involves a different part of the brain. This helps to explain why multilingual patients who have had a small stroke or suffered localised brain damage through surgery may selectively lose only one of their languages.

Learning to Sing

The learning of songs by birds is in some respects analogous to the acquisition of language in humans. Like humans, song birds start the process early in life and normally copy the songs of their own species. The typical pattern is for the young male bird to listen to and store sounds made by his father and other males during the first few months after he has hatched.[17] The following spring he produces a range of sounds and, by degrees, settles on songs he has heard before. When he is mature, he uses his songs to defend territory and attract females. The range of songs acquired by each male is transmitted within the neighbourhood from one generation to the next, in much the same way as language, customs and ideas are transmitted across the generations in humans.

The song-learning ability in birds is linked to a specific part of the brain. Learning is accompanied by the growth of new neurons and by changes in connections between neurons in certain brain areas. Some brain sites are needed for the capacity to produce song at all times. Others, particularly a region of the forebrain called the lateral magnocellular nucleus, or lMAN, are only needed for the process of *acquiring* song, and seem to perform no function in producing the song once it has been learned. It is possible in experiments to damage small areas of the brain. Damage to the lMAN of zebra finches early in life disrupts the development of song. But if the brain lesions are made later in life, they have no effect on how well the birds sing.[18] In this sense, the lMAN is like the neural scaffolding that has been suggested for human language development.

The image of temporary neural scaffolding, used for assembling bird song or human language, is compelling. But sensitive periods in development may arise in other ways as well. Preferences and habits, once in place, may make the formation of other preferences and habits more difficult because of disinclination rather than lack of ability.

Closing Options

The more you get used to riding a bicycle with upright handlebars, the more difficulty you will have riding one with dropped handlebars. The more you have roast chicken, the less you will like roast maggots. Some people are suspicious of novel food, and the more restricted their early experience the more they will tend to reject food that is unfamiliar in colour, texture or taste. Even during terrible famines, people have been known to refuse food simply because it was strange. In the Bengal famine of 1943, in which more than a million people died, efforts to persuade hungry people to eat relief supplies of wheat and millet met with remarkably little success. In Bengal the diet of the poorer people consisted largely of rice. The domestic baking of bread was unknown and the making of light and palatable unleavened bread, or chapatti, required an iron grid and considerable skill. The poor had neither. An additional problem was that, as rice-eaters, they were accustomed to bulky meals of soft consistency and did not relish more concentrated foods such as chapatti, which differed in consistency and needed more chewing. Despite starvation, these people were reluctant to switch to a new type of food. Campaigns to persuade them to 'Eat more wheat' were generally ineffective. The report of the official inquiry into the causes of the famine acknowledged that weaning rice-eaters on to alternative grains had proved difficult.[19] In a similar vein George Crabbe wrote in *The Borough*, his poem about the people of Aldeburgh on the east coast of England, 'Habit with him was all the test of truth, "It must be right: I've done it from my youth."'

The more ingrained a preference or a habit becomes, the lower the chances of a radically new preference or habit being adopted later in life. The narrowing effects of experience typically make people more conservative as they age. Forming preferences and habits has a pre-emptive effect, shutting out exposure to novelty and to new ways of performing old tasks.

The individual changes in such a way that further change becomes more difficult. (It may not be impossible, however, and in Chapter 10 we shall describe ways in which habits of a lifetime may break down.)

Behavioural changes are mirrored in the brain. Structures in the brain that have been altered by earlier experience may pre-empt the formation of new structures. Permanent poor sight, known as amblyopia, sometimes resides in the brain itself, rather than in the eye, and derives from unusual visual experience early in life. The development of a child's binocular vision is seriously affected if he or she had a squint in the first few years of life. The child becomes dependent on one eye and the other eye becomes virtually useless. The squint can be corrected by surgery, in which case the child usually develops normal binocular vision, but only if the surgery is performed sufficiently early in development. If it is left until much after the age of three years, surgery will do no good and the child is left with permanently impaired vision in the affected eye.[20]

The brain mechanisms involved in visual development have been analysed in cats and monkeys. The capacity of an eye to activate neurons in the cat's visual cortex depends on whether that eye received visual input during the first three months after birth.[21] If one eye is visually deprived during this period it virtually loses its capacity to excite cortical neurons. The other eye then becomes dominant and, once established, usually remains dominant for the rest of the individual's life. Much the same is true in humans.

Young men sometimes contend, self-servingly and without justification, that without frequent sex they will lose their virility, if not their actual penis. In the case of the developing nervous system the adage 'use it or lose it' comes closer to being true. In the cat brain the changes needed to establish the dominance of the used eye over the unused eye take place in a particular layer of the visual cortex. It appears that, once one set of neurons has established a connection, they exclude others from doing so thereafter. Special neurobiological mechanisms, involving particular receptors at the sites of plasticity, are required for such pre-emptive changes to occur. Once a stable pattern of responding has developed, these mechanisms fall away.

If the brain is not stimulated in the normal way through the visual pathways then the reduction in the number of these receptors is delayed until the necessary inter-connections between neurons have been established.[22] It is tempting to suppose that these receptors provide the scaffolding for the construction of a stable visual pathway. If the visual pathway has not been constructed for some reason then the scaffolding stays in place. Thus, if the cat is deprived of normal visual experience during the first few months of life, the sensitive period during which neural changes can occur is extended. Sensitive periods are not infinitely flexible, however. If the lack of visual experience goes on long enough, the sensitive period will eventually terminate and the animal will end up with a brain that differs from the normal. The train eventually leaves the station.

Neural machinery that is normally used for one purpose may be redeployed if the necessary sensory input is absent during early development. Brain imaging has shown that the area of the brain that normally processes visual information becomes reorganised in blind people. In individuals who have been blind from an early age, tactile cues stimulate parts of the primary visual cortex.[23] This does not happen in sighted people. The change in brain organisation was revealed by an experiment in which strong magnetic fields were used to disrupt the function of different cortical areas in people who had been blind from an early age. Disruption of the visual cortex disrupted their ability to read Braille or embossed letters. In sighted people, on the other hand, transient disruption of the visual cortex had no effect on their ability to perform tactile tasks. Evidently the brains of the blind people had been reorganised in response to their particular experience of the outside world.

The physiological analysis of developmental abnormalities in the mammalian visual system has yielded valuable insights into what happens under normal conditions. Even so, the times of peak activity in the special-purpose learning processes raise issues of greatest interest when examining the development of behaviour. Analysis of how they work reveals that relatively simple underlying processes can generate great behavioural complexity. One of the best illustrations is provided by a remarkable process of social attachment in birds and mammals, known as behavioural imprinting.

Stamping in

Behavioural imprinting is the process by which a young animal, guided by certain predispositions, rapidly learns the details of its mother's individual appearance and forms a social attachment to her.[24] (Confusingly, the term 'imprinting' has also been used for a quite different process by which a parent may influence gene expression in its offspring.) Some young animals also learn to recognise their father as well, but this is less common since, particularly in mammals, fathers rarely play a substantial role in caring for their offspring.

In the natural environment behavioural imprinting reliably results in the formation of a strong social bond between offspring and parent. The parent must recognise the offspring in order not to waste time and energy caring for the young of others. The offspring must recognise its parent because it might be attacked and even killed by other adults of the same species which do not recognise it as their own. The rapid learning process is such that, if the young animal is reared under abnormal conditions, imprinting can result in the formation of bizarre attachments. Ducklings or domestic chicks which have been hand-reared for the first few days after hatching strongly prefer the company of their human 'parent' to that of their own species. The pictures of Konrad Lorenz being followed around by goslings, or arching their necks over his head while he swam in an Austrian lake, are as famous as the term he used to describe the learning process.

The word 'imprinting' suggests that a permanent irremovable image has been left by the impact of experience on the soft wax of the developing brain. According to the original idea of a sensitive period for learning, the brain's metaphorical wax is soft only during a particular stage in development and no impression can be left before or after the 'critical' period. However, subsequent research has revealed that the image of such a sharply delineated moment of imprinting is misleading, because the process is not so rigidly timed and may indeed be undone. This is why the term sensitive period is now most commonly used to refer to the phase during early development when the young animal most readily forms a social attachment.

Behavioural imprinting seems to have been well designed for its biological function. The onset of the sensitive period for imprinting is accompanied by other developmental changes that make it much easier for the young animal to learn about the details of its mother's physical characteristics. The young animal's vision improves at around this time and it begins to move about with greater ease. The timing of this cluster of developmental changes depends on the species in question.[25] In birds that are hatched blind, naked and helpless, such as swallows, the onset of imprinting occurs much later in relation to hatching than it does in the precocious ducklings, which are already feathered and active when they hatch. Changes in the animal's abilities to perceive and deal with the external world play an important role in determining when a sensitive period starts. These abilities must themselves develop, and the rate at which they do so may depend on the animal's experience.

Competitive Exclusion

How does behavioural imprinting work? It has often been pictured as though a developmental program, linked to an internal clock, opens a window onto the external world, allowing the animal's behaviour to be altered by experience. After a predetermined time the program was thought to shut the window and thereby remove the possibilities for further modification of behaviour. These images of internal clocks, programs and windows onto the environment offer an attractively simple model of how imprinting might work. But they are not entirely satisfactory.

First of all, the imprinting process is not so sharply defined as the image of windows opening and shutting implies. More importantly, the images of clocks and windows do not offer a real explanation of the biological mechanisms underlying sensitive periods. The observational evidence simply shows that at a given stage in development the individual's subsequent development can be modified by experience. But this evidence is not in itself an explanation for the ineffectiveness of experience outside that sensitive period. Nothing new is added by stating that the experience has to occur within the sensitive period in order to be effective, since that is precisely the evidence that requires explanation.

The timing of behavioural imprinting can, within limits, be adapted to circumstances. Exposure to one object leads the individual to prefer it and reject anything perceived as different. Thus, experience of one sort prevents other types of experience from having the same impact. Such competitive exclusion is similar to what happens to visual development when one eye is covered up in early life. If the developing animal is deprived of the experience it would normally receive during the sensitive period, the imprinting process slows down and the sensitive period is lengthened. Hence the sensitive period for imprinting is extended in domestic chicks if they are reared in isolation from other chicks. But rearing the animals in isolation does not simply delay a hypothetical internal clock from bringing the sensitive period to a close. It eventually leads to the formation of a preference for whatever the individual has experienced, no matter how unnatural. Accordingly, chicks which have been reared in social isolation in a pen with patterned walls will eventually form a memory for that pattern and will subsequently respond socially to a moving object if it bears the same pattern.[25] This finding is important because it shows that sensory deprivation during the normal sensitive period for imprinting does not stop the nervous system from slowly settling into a highly abnormal organisation. Despite its abnormality, the outcome reflects the same process that would lead under normal circumstances to the successful development of a social preference. In the natural environment the bird is virtually certain to be exposed to its mother sooner or later.

This flexibility of the imprinting mechanism is important because it allows for naturally occurring variations in conditions. When the weather is warm, for example, the mother duck will lead her ducklings away from the nest within hours of hatching. When the weather is cold, however, the mother will brood her young for several days after they have hatched and the young may consequently see little of her because they are underneath her. The neural machinery responsible for imprinting has to be sufficiently flexible to cope with this variability in the young animal's experience.

The neural mechanisms underlying the imprinting process and sensitive periods have been analysed in the domestic chick in a long series of studies by Gabriel Horn and his collaborators at Cambridge University and elsewhere.[26] A central problem in the study of imprinting has been finding where in the brain the information about the mother (or her substitute imprinting object) is stored. An array of neurobiological techniques has implicated one particular region of the chick forebrain as the site where the neural representation of the imprinting object resides. This site is called the intermediate and medial part of the hyperstriatum ventrale, or IMHV.

How was the neural seat of imprinting located? It is not good enough simply to show that a particular part of the brain is active when the bird is learning about the imprinting object. This is because lots of other things happen during the imprinting process: the young bird is visually stimulated and aroused by the imprinting object (normally its mother) and it also tries to approach and follow the object. All these processes produce their own changes in brain activity. When experimental evidence is open to a variety of different interpretations, greater confidence in one particular explanation can be attained by tackling the problem from a number of different angles. The experiments applied to imprinting are worth recounting in some detail because they illustrate how behavioural aspects of development can be linked to an understanding of the brain mechanisms.

The first approach took advantage of the fact that, in birds, all the sensory input to the brain from one eye can be restricted to one hemisphere of the brain by cutting a bundle of nerve fibres running between the two hemispheres. After this had been done, one of the chick's eyes was covered with a patch, so that it could only see the imprinting object (a flashing rotating light) through one eye. This procedure meant that only one side of the chick's brain was exposed to sensory information about the imprinting object. When this was done, a difference in brain activity between the exposed and unexposed sides of the chick's brain was found only in the forebrain roof. No differences between the two sides were observed in other regions of the brain. This 'split brain' technique eliminated the possibility that both sides of the brain were affected equally by training.[27] However, it did not exclude the possibility that the enhanced brain activity was due to greater visual stimulation of the trained side. Other procedures were therefore needed.

Another set of experiments exploited the fact that individual chicks differ naturally from one another in their behavioural response to imprinting. Various aspects of the chicks' behaviour were measured while the chicks were being trained, and their preferences for the familiar object were then tested. This procedure opened up to examination the relationships between behavioural measures of imprinting and neural activity in different parts of the brain. Only one behavioural measure was positively correlated with biochemical activity in the roof of the anterior forebrain – namely, how much the chicks preferred the familiar object to a novel object when given a choice between the two. This index of learning was not correlated with biochemical activity in any other region of the brain and, equally important, was only weakly linked with other behavioural measures such as the birds' overall behavioural activity and responsiveness.[28] The analysis therefore revealed a specific link between a behavioural measure of imprinting and biochemical activity in a part of the brain that had already been identified in other experiments as the seat of imprinting.

In further experiments, the amount of imprinting experience the chicks received on the first day after hatching was varied. Some chicks (the 'under-trained' group) received only a limited opportunity for learning about the imprinting object on day one, while others (the 'over-trained' group) were imprinted for so long that they could learn little more about the imprinting object. On the second day of the experiment, all the birds were exposed to the same object for the same amount of time. If biochemical activity in the forebrain roof is specifically related to learning, then birds that had previously been over-trained should learn less about the imprinting object on the second day, and consequently should show less activity in the forebrain roof than the under-trained birds. The results supported this. As the length of exposure on the first day increased, so the biochemical activity associated with changes in the forebrain roof decreased on the second day. No such relationship between learning and brain activity was found in any other brain regions.[29]

Collectively, these converging sets of experiments established that increased biochemical activity in the chick's forebrain roof is necessary for, and exclusively related to, the storage of information after imprinting. None of the experiments by itself ruled out all the alternative explanations. Each piece of evidence obtained by the different approaches was ambiguous by itself, but the ambiguities were different in each case. Therefore, when the whole body of evidence is considered, much greater confidence may be placed on the final interpretation. An analogy is the process of triangulation – locating on a map the position of a mountain top. One compass bearing is usually not enough; two bearings from different angles provide a much better fix; and three bearings give a reliable position.

Careful mapping of the chick brain has confirmed that a neural representation of what is learned during imprinting is stored permanently in the IMHV region.[26] Chicks that have had both their left and right IMHVs removed surgically are unable to imprint. Moreover, if both IMHVs are damaged experimentally immediately after imprinting has taken place, the birds show no recognition of the imprinting object. Damaging the IMHVs erases their memory for the imprinting object. These experiments have demonstrated that the IMHV is needed both for imprinting to take place and for the memory of the imprinting object to be recalled after imprinting has occurred. Further studies have revealed that another neural representation is formed or confirmed in a different region of the brain about six hours after imprinting. This representation can be prevented from forming by surgically damaging the right IMHV soon after imprinting. In the critical IMHV regions, connections between neurons are both enlarged and diminished. The enlarged neural connections are specific to features of the particular object with which the bird has been imprinted, and the diminished connections are specific to features that are not present in that object.[30] These discoveries strongly support the competitive exclusion hypothesis for imprinting because they show that exposure to one imprinting object makes another imprinting object less effective, and these changes are reflected in the chick's brain.

The understanding of the neural mechanisms of behavioural imprinting that has accrued from studies such as these does not, of course, mean that all sensitive periods work in the same way. But it does show how relatively simple mechanisms can satisfy a complex and biologically important requirement for an animal. The developmental process is well designed to gather important information about a key individual (the mother) and store that information for later use. Moreover, the stored representation is well protected from change after the imprinting process is complete.

Learning the Lessons

A single, all-encompassing theory of sensitive periods in development is an illusory goal, simply because the range of phenomena covered by the term is so diverse. Some sensitive periods merely represent a stage in development when the young individual is particularly vulnerable to disruption. Other sensitive periods have evolved for specific reasons – usually because of the need to acquire certain sorts of information about parents or close kin at the right time in development. These different types of sensitive period undoubtedly rely on different biological mechanisms. The processes that determine when in childhood the acquisition of language occurs, for example, are bound to be different from those that affect what shape and size of body the child will build.

A number of important conclusions have emerged from studies of sensitive periods for learning. The developmental processes that make learning easier at the beginning of a sensitive period are often linked to physical growth, and they are timed to correspond with changes in the habitat of the developing individual. The processes that bring the sensitive period to an end are related to the gathering of crucial information and, except in extreme cases, do not shut down until that information has been gathered. The ending of the sensitive period must reflect the variable opportunities for gathering that information in the real world. To use the building analogy, the scaffolding is not dismantled if delays in supplying the raw materials have meant that the building is incomplete. A limit is set, however, because so much else has to be done in development. If the relevant information remains unavailable for too long, the individual may eventually have to make the best of a bad job and move on.

Although the use of a common term like 'sensitive period' should not imply a common underlying mechanism, it does draw attention to the basic fact that all forms of experience are not equally important at all stages in development. Some scientists have treated the sensitive period as an explanatory concept rather than a descriptive term, and have reached enthusiastically for metaphors involving a developmental 'window' opening and then shutting again. But the evidence from real organisms reveals a dynamic interplay between the individual's internal organisation and the external conditions. The actual chemistry of sensitive periods does not look at all like a computer program.

Images and metaphors do have their place, however, and the building analogy is helpful when considering how sensitive periods might work. It suggests that special-purpose devices, like scaffolding, are needed when an elaborate structure is under construction. These devices do not have to remain in place for ever, just as scaffolding is eventually removed from a building. If for some reason the raw materials are not delivered on time and the building site is valuable, then other structures may start to spill over onto the vacant spot, or it may eventually be used for another purpose altogether.

Pursuing this analogy, buildings are solid objects, and mortar and concrete set hard. It is equally true that a body, once built, is difficult to alter. Making fundamental changes to mature behaviour patterns or personality traits will similarly take time, resources and quite possibly support from others. Adults have important tasks to carry out, such as feeding and caring for their family, and cannot readily dissolve themselves and reconstruct their behaviour without others to care for them during the transition phase. And if the developing brain is regarded not as a building, but as a wax model, then a different prospect is raised. Experience adds pieces to the model and sculpts others away, generating a form that may harden and be difficult to alter. But warm up the model and change is once again easy. This image has important implications for the understanding of human development. In Chapters 9 and 10 we shall consider the circumstances in which both continuity and change are found in human development.

9
Morning Shows the Day

The childhood shows the man,
As morning shows the day.
John Milton, *Paradise Regained* (1671)

Life Sentence

Gardeners like to get their plants off to a good start, pinching out buds to form an even spread of a shrub's branches, or staking a tree to encourage straight growth. If on the other hand asymmetry is the prized quality, as in some miniature bonsai, the unfortunate plant is wired into a romantically skewed position. 'As the twig is bent, the tree's inclined,' wrote Alexander Pope. The physical constraints placed on the growing plant early in its development determine its adult form.

Obsessed as they were with the image of shaping their children's growth and posture, nineteenth-century middle-class European parents sometimes strapped their offspring into iron frames. The discomfort brought no gain and the market for these devices eventually collapsed. Nowadays, the astonishing range of gleaming hardware to be seen in the mouths of teenagers testifies to the fact that teeth, at least, are malleable. If you pay enough money, your child can have a film star's smile. Can behaviour also be pruned, directed and confined so that the adult follows the pattern shaped in the child? The short answer is 'yes – sometimes', but the story has many strands.

The existence of sensitive periods suggests that, for various reasons, events occurring early in life may leave a more lasting impact than those occurring later on. Some of these effects arise because the young are especially vulnerable at times of rapid growth or change. Others arise because individuals act on the basis of environmental forecasts, setting in train one of several alternative patterns of life that are not easily unravelled once established. Information that is essential for a fully functional life may be pre-emptive, blocking later-arriving information from exerting a comparable influence on preferences or habits. One bad experience may mean that the individual subsequently does not risk investigating whether circumstances have changed. This is particularly noticeable in food aversions. Perhaps you were once violently sick after eating food that was new to you; if so, you probably did not try that food again for a long time. The aversion can be so powerful that you may not have lost it even if you know rationally that the sickness was coincidental. Simone de Beauvoir recorded these views on the theme of continuity:

> I have also noted a great stability in what is called people's character – their reactions, taken as a whole, in analogous circumstances. The passing years bring changes in the individual's situation, and this affects his behaviour. I have seen shy or sullen adolescent girls become happy young women in full flower . . . But generally speaking men and women, once they are settled into adulthood, remain consistent. Sometimes, indeed, they repeat their own conduct when they think they are being quite different.

Buried in her observations are two points. First, once formed by the cooking processes of early development, individuals are set into patterns of behaviour which they are likely to repeat throughout their lives. Second, individuals may differ from each other in ways that are present from birth and persist despite great similarities in background. Continuities are found across many or all of the seven ages.

Wiping the Tapes?

Passing through the seven ages of development involves substantial reorganisation of the body and behaviour. Birth brings the biggest changes, but those involved at puberty are nearly as dramatic. These major discontinuities during development are reminiscent in some respects of the dramatic process of metamorphosis, whereby a flightless caterpillar is transformed into a butterfly, or a water-bound tadpole turns into a toad. The transitions during the human lifespan have provoked psychologists to ask whether memories stored at an earlier stage of development are subsequently 'erased from the tape', to use Jerome Kagan's phrase.[1] Undoubtedly, much of what a child has learned before the age of two is later forgotten. But if that is so, how do the effects of early experience persist into later life? One possibility is that continuity from one stage of an individual's development to another is ensured by continuities in their environment. The conditions of the environment that caused a child to behave in a particular way may persist, even though the child has gone through a period of change. The child may simply relearn what it had learned before and, as a result, behaves as it did before its metamorphosis.

Even in the animal examples of metamorphosis, however, the image of 'wiping the tapes' is too extreme. In insects and amphibians, some effects of an individual's early experiences persist even after metamorphosis has occurred.[2] For example, adult female moths tend to lay their eggs on the particular type of food on which they were reared as caterpillars. The memory survives metamorphosis. In amphibians, the effects of conditioning during the larval phase can last into adulthood. Likewise, mammals retain memories from early in life. In one experiment rats were conditioned during the first month after birth, using intermittent sleep interruption paired with a sound. When the same sound that had previously signalled interrupted sleep was presented to those individuals as adults, they recognised the sound and responded physiologically.[3] In humans the effects of an unusual experience at six months of age may be detected two years later, when much else has changed within the child.[4]

The process in human development that probably comes closest to metamorphosis, in terms of the scale and abruptness of change, is birth. The possibility that pre-natal experience may influence the growth and development of the brain has provided a commercial spur for new markets. Parents on the lookout for ways of boosting their children's intelligence are nowadays tempted with electronic gadgets that claim to do just that. One company, for example, exhorts pregnant women to wear a Walkman-like device around their waists during the final four months of pregnancy. The gadget produces thumping and whooshing sounds which broadly mimic the mother's heartbeat but differ subtly from the real sound. The foetus is supposed to notice the difference and is supposedly aroused by it. The idea behind the device is that this stimulation accelerates the growth of the foetal brain. Babies born to mothers who have used the device are said to be more alert, more supple and to have better muscle control. Whilst nobody has suggested that such devices are likely to do any harm, hard scientific evidence for any beneficial long-term effects is elusive. On the other hand, we have already described in Chapter 6 ways in which pre-natal experience exerts long-term effects on people's bodies and behaviour. Some of these effects of pre-natal experience are mediated by changes in metabolism, patterns of growth and responses to stress, and do not require the survival of memories.

But we also described in Chapter 2 how the particular sounds of the mother's voice, experienced before birth, are recognised after birth. So, such metaphorical wiping of the tapes as occurs, is only partial. Continuities can be found.[5]

More than a Seed

Temperamental or behavioural characteristics such as shyness or bad temper are particularly likely to remain unchanged over long periods of an individual's lifespan, for a variety of reasons.[6] The issues surrounding continuity are illustrated by studies of the child's emotional attachment to its mother. In the aftermath of the Second World War, the World Health Organisation was concerned about the long-term mental health of children who had been orphaned or separated from their families. John Bowlby, an eminent British psychiatrist, was commissioned to study the problem. In 1951 he published a seminal monograph, which continues to excite discussion to the present day. Bowlby found that socially disruptive adolescents were more likely than others to have been separated from their mothers. He argued that 'the prolonged deprivation of the young child of maternal care may have grave and far-reaching effects on his character and so on the whole of his future life'.

The links that Bowlby described between an adverse experience early in life – separation from the mother – and behavioural problems later on might have arisen in less obvious ways. The affected children might, for example, have had difficult personalities from birth and been instrumental themselves in their early deprivation. Their persistent crying and bad behaviour might have caused a breakdown in the health of the mother, resulting in separation. According to this argument, the seeds of these adolescents' behavioural problems were present before they became separated from their mothers. While not wholly implausible, this interpretation runs counter to most of the evidence, which instead supports the view that the children's adverse experience is normally the cause rather than the consequence of their behavioural problems.[7]

Bowlby's pioneering study of maternal deprivation caused a furore because it was misinterpreted as suggesting that separation from the mother or principal care-giver results in prolonged emotional disturbances in the child which persist into later life. The implied advice for mothers was that they should remain with their children throughout the formative years from two to five. This was not well received in the 1960s and 1970s, when many women were trying to break away from the constraints of tradition and combine child-rearing with professional careers. Nor was this interpretation of Bowlby's work correct.

Bowlby's study looked backwards. It relied on finding people who had become delinquent as adolescents or adults, and then inspecting what had previously happened to them in childhood. Bowlby found that most individuals with problems in adolescence or adulthood had had a bad childhood. But forward-looking research has painted a different picture. When a large number of children are tracked from childhood onwards, the maternal deprivation that some of them endure seems to have little or no long-term effect in many cases. The presence of others in the family, who care for the child while the mother is absent, is often crucial. Just before he died, Bowlby looked back on the field of research which he had spawned.[8] He clearly recognised the variety of influences that maintain the effects of early separation. So the child at greatest risk is the one who was separated from a single mother with little family support. A particular experience, just like a gene, may only produce a noticeable effect on the individual in certain circumstances, while under other conditions it leaves no lasting impression.

The chain from separation to subsequent behavioural problems is long, with many links in it. For example, breakdown of parental care may mean that the child ends up in an institution. On leaving the institution years later, the adolescent returns to a discordant family environment. These stresses encourage the adolescent to 'escape' into an early marriage. Poor choice of partner, the immaturity of the couple, and an early pregnancy then contribute to marital breakdown or to an unhappy, unsupportive relationship. The tensions lead to collapse of parental care for their children and the vicious cycle starts all over again.[7]

What about personality characteristics such as shyness? Again, the mechanisms may be subtle. Psychological studies of shyness in children have shown that shyness and social withdrawal typically remain stable over time.[6] They also suggest that reserved children may lock themselves out of opportunities for close contact with other children and hence, without the normal opportunities for socialising, their shy temperaments stay unchanged. Their shy behaviour may also evoke responses from other people which tend to reinforce and sustain that behaviour. Both processes play a role in maintaining the continuity in characteristics such as shyness and bad temper. They are sustained because the individual is channelled into environments – especially social environments – which tend to reinforce the original behaviour.

Infants who are easily distressed and agitated by unfamiliarity at the age of four months tend to become shy and subdued in early childhood. Conversely, infants who are equable at four months are more likely to become bold and sociable later on.[9] The differences might be due to genetic differences. Indeed, studies of rhesus monkeys, in which comparable personality differences are found, have suggested that genetic influences are present, since offspring resemble their fathers, who play no part in rearing the young.[10] Nevertheless, personality characteristics established in early life as a result of social deprivation are maintained despite big changes in the social and physical environment.[11] The evidence suggests that, in humans too, the mother's behavioural style has a direct and long-term effect on the characteristics of her child.[12]

Staying on Track

Some children are psychologically much more resilient than others. Whereas some seem scarcely affected by terrible experiences, others are psychologically scarred for life. The American psychologist Emmy Werner pioneered the study of developmental resilience in the 1950s. Her idea was to focus on those children who manage to cope with adversity. What is distinctive about them? Werner conducted a long-term study of development among a multi-racial cohort of children born in 1955 on Kauai, one of the islands of Hawaii. Many of the children came from poor backgrounds, with alcoholic, abusive or even psychotic parents. They had suffered stress before birth and much family discord afterwards. Their psychological and social development was assessed periodically over the next forty years. Werner found to her surprise that many of the children who seemed so much at risk early in their lives nonetheless developed into competent and confident adults. About a third of them – the 'resilient kids' – coped remarkably well despite their traumatic upbringing. They had certain psychological characteristics that may have helped them cope. In particular, those who coped well tended to be planners and problem-solvers. Their intelligence and cheerfulness probably helped to circumvent problems that would otherwise have arisen.[13]

Children who have suffered appalling neglect are capable of showing remarkable recovery in their cognitive and social development – if they are rescued sufficiently early in development. Children who have been isolated in cupboards, attics or cellars exhibit major learning difficulties when they are found. Yet they usually recover to normal levels of intellectual functioning and language development if they are rescued by the age of four or five. Given sufficient care, even children who are not rescued until adolescence can still be helped a lot. One group of terribly deprived Romanian fifteen-year-old orphans were adopted into Canadian families. On adoption, they were substantially below average in body size and had many behavioural problems. Abnormal behaviour persisted in a third of the children when they were followed up three years later. But remarkable improvements had occurred in the majority.[14]

An extraordinary case of developmental resilience was that of the blind and deaf American writer Helen Keller, who died in 1968 at the age of eighty-eight. Keller lost her sight and hearing following a major illness at the age of nineteen months. Soon afterwards she also became mute. When she was about six years old she started to receive special tuition from Anne Mansfield Sullivan, a graduate of the Perkins School for the Blind in Boston. Sullivan herself had been blind, but had partially recovered. With the help of Sullivan's intense and specialised teaching, together with instruction at special schools for the deaf and blind in Boston and New York, Keller learned to read and write in Braille. She also learned to speak by pressing her fingers against Sullivan's larynx to feel the vibrations. In 1904 Helen Keller graduated with distinction from Radcliffe College. She devoted her life to promoting the interests and education of the deaf and the blind through her writing and world tours. Her voice remained difficult to understand and she required the service of a 'translator' when speaking to public audiences. Keller's triumph over her disabilities was remarkable, but it would not have happened if she had not received intensive training and assistance during her early years.[15] Helen Keller's story emphasises that resilience occurs in a social context. The child is helped by other people – usually parents and teachers.

This social aspect of resilience is found in other species as well. Environmental pollutants such as lead can seriously affect the developing brains of humans and other animals. If one of a pair of herring gull chicks is injected with a lead compound shortly after hatching, the chick's behaviour is disrupted; it is consequently less able to compete for food with its siblings and gains weight less rapidly. However, the parent gulls, both of whom care for their young, are able to perceive these differences in capability between their offspring and compensate for them in a remarkable way. After one parent has started to feed the larger chick – the one which was not impaired by lead – the other parent moves a short distance away and calls to its impaired sibling. When the second, smaller chick approaches, it is fed. This compensatory extra parental care results in the lead-impaired chick growing more rapidly and catching up to its sibling's weight by the time of independence. The lead-impaired chick displays developmental resilience in the face of an initial disadvantage, but this resilience operates mainly through the mechanism of its parents' behaviour.[16]

A comparable example of parental compensation in mammals was discovered by Peter and Martha Klopfer. Peter Klopfer is a zoologist who works at Duke University in North Carolina and lives on a nearby farm. Among other animals, he and his wife keep Toggenberg goats, which regularly produce twins. The twins usually differ in size at birth. The Klopfers noticed that, provided the smaller twin is sufficiently vigorous, the mother inhibits the bigger one from suckling by vigorously licking it. The mother's behaviour allows the smaller twin to get more milk; it consequently grows more rapidly until the twins are of the same size. Again, parental compensation evens out an initial difference between offspring.[17]

Developmental resilience can also operate without the help of parents. If a child has been seriously ill for a while, its growth may have slowed or even stopped. Then, after the illness is over, growth accelerates until the child's height returns to where it would have been if illness had not intervened. The child catches up. The same catch-up growth occurs in children and young animals of other species after a period of starvation. If a juvenile rat is starved, its weight drops; but when it is put back onto a normal diet its growth rate rapidly picks up, returning its weight to that of an unstarved rat. The body seems to know how big it should be at a given age and, with natural elasticity, bounces back.[18]

Conrad Waddington was one of the most imaginative thinkers about development in the middle of the twentieth century, drawing together the ideas of many of his contemporaries. In a delightful visual image, Waddington represented development as a ball rolling down a tilted plane which is increasingly furrowed by numerous diverging valleys. The ball represents the developing individual or any one of its characteristics, and the valley down which it rolls is part of what Waddington called the 'epigenetic landscape'. As the ball rolls into an ever-deepening valley, its sideways movements become increasingly restricted, or 'canalised' as Waddington termed it. If the ball encounters an obstacle and is not stopped dead, it rides up round the obstacle and falls back into the valley down which it had been rolling. This attractive image provides a useful way of thinking about different developmental pathways, and the capacity of the developing system to right itself after a perturbation and return to its former path.[19]

Using a computer it is relatively simple to simulate a growing organism that can compensate for short periods of food deprivation during its development. The amount of food the individual attempts to eat is simply determined by a comparison between a fixed setting – the preferred size – and the actual size at that time; the value of the preferred size is then increased as the individual ages. Simple mechanisms such as this can result in different developmental routes leading to the same end state. Some of the characteristics of development which in an earlier age seemed to suggest vital forces, do not in fact pose huge conceptual problems. However, other aspects of development do look as though rather more complicated explanations are required.

Different Routes to the Same Place

Weary commuters arriving at the station to catch their train home sometimes find that the trains are not running. If they are sufficiently determined they will usually find an alternative way home, using another train route, bus, taxi, private transport, their feet, or a combination of these. An analogous process of switching to alternative routes is found in the development of bodies and brains. Like the weary commuter, the developing individual eventually arrives at the same destination but gets there in an entirely different way.

Systems theorists writing in the middle of the twentieth century laid considerable emphasis on the self-correcting features of development. They produced theoretical models showing how developing systems could reach the same end state via a variety of different routes. This convergence of different developmental routes on the same final end state was referred to as 'equifinality'. While it is easy to become a little mystical about concepts such as equifinality – and some systems theorists did not always resist that temptation – it does at least highlight the remarkable elasticity of development. Nowadays the properties of inanimate dynamical systems have been used to explain the way in which many different routes can lead to a stable end point.[20]

During the early stages of development the brain has the capacity to reorganise itself in remarkably different ways. Consider, for example, the pre-frontal lobes of the human brain, which play an important role in deciding which one of many incompatible things an individual may do. In Chapter 7 we described how people whose pre-frontal lobes are destroyed by injury or disease inhabit an emotionally flat landscape in which it is difficult for them to make decisions even though they are perfectly capable of solving a problem that requires a rational solution.[21] Experimental studies have found that rhesus monkeys with the same sort of damage to the pre-frontal lobes are similarly unable to solve behavioural tasks in which they have to decide between alternative courses of action. However, if the same brain damage is inflicted before birth, the monkeys' decision-making capacity is unimpaired. When the brains of these monkeys are examined after they have died, the anatomy is quite different from that of a normal monkey; indeed, the appearance of each brain is so unusual that it might have come from another species. In the face of a major disruption, brain development was abnormal, but behaviour developed as if nothing had been done to the animals.[22]

When parts of the brain are damaged early in life, the functions that would normally be carried out in that region are sometimes transferred elsewhere in the brain. Hence brain scans occasionally uncover people who evidently received minor brain damage earlier in their lives, probably before or during birth, but whose behaviour has been apparently normal. The brain scan is often the first and only evidence of any problem. One of the most startling examples was that of a 26-year-old man who had received a first-class honours degree in mathematics from a British university. He was of above-average intelligence, with an IQ score of 126, and in most respects behaved quite normally. However, he had an unusually large head and his movements were mildly unco-ordinated. It was these mild abnormalities that brought him to the attention of neurologists. A brain scan showed that, instead of the usual 4.5 cm-thick layer of brain tissue between the fluid-filled ventricles and the surface of the cortex, this man had a cortex that was only a few millimetres thick. The contents of his skull were mostly cerebrospinal fluid, resulting from hydrocephaly in early life. But despite this apparently severe distortion, his brain had apparently reorganised itself and most behavioural functions had developed normally.[23] Since this case was discovered in the late 1970s, many comparable examples have been found of individuals with a grossly abnormal brain structure yet normal mental faculties. They constitute striking instances of equifinality and normal functioning against all odds.

An engineer designing a robot to cope with unpredictable conditions will focus on the machine's purpose. So long as that ultimate goal is reached, the precise means by which it is achieved are of secondary importance. The robot will therefore be designed to keep acting according to pre-arranged rules until sensory input from the environment signals the attainment of its goal, triggering a cessation of action. An early example was Gray Walter's battery-run robot tortoise, which needed periodically to return to an electrical socket to recharge its batteries. It was designed to keep moving until it could lock into a signal from the socket and approach the place where it would able to 'feed'.

Analogous mechanisms in the developing brain may be responsible for its astonishing resilience. As the brain is formed, populations of neurons migrate from one region to another or establish remote connections. Long, thin outgrowths of 'pioneer' neurons pass between many other cells until they reach their target in what may be a distant region of the brain.[24] In the same way as the robot, the migrating cells or their growing tips may use simple search rules to reach their specific goal. If they encounter an obstacle, such as a region of damaged brain tissue, the neurons bypass the obstacle and take an unusual route. Provided the neurons eventually end up in the right place, brain function may be normal, even though its architecture is dramatically different. Only a careful examination of the brain as it develops will reveal the unusual way in which normal functioning has been attained.

The Strands of Continuity

The strongly held belief of those who were steeped in eighteenth-century Enlightenment and of the Romantics who followed them was that, in William Wordsworth's phrase, 'The child is father of the Man.' The constancy of personality and behavioural characteristics can generate a gloomy sense of fatalism when life seems to have run into a cul-de-sac. Surely it is possible to escape? Back comes the response that once you are locked into a particular way of behaving, nothing can be done to change it. While resilience is an important facet of development, it is far from obvious why some people cope better with adversity than others. Many cannot cope and are left with effects that damage the rest of their lives. Rehabilitation is as important as prevention, since continuities of a person's characteristics across time are often due to the persistence of the same social conditions. Somebody who experiences an adverse environment early in life often continues to experience adversity later in development as well; children from poor homes usually end up in poor schools, mix with other poor children and get poor jobs. Continuities reside both in the environment and in the individual.

Environmentalist views are countered by observations that personality characteristics often run in families and identical twins are more alike than non-identical twins. But recent evidence has also shown how the social environment influences the long-lasting characteristics of an individual's personality. Early relationships are important in establishing an individual's style of dealing with others. Here, as elsewhere, many ingredients play their part. Once cooked, though, a person's characteristics may remain the same for a long time.

10

Room 101

> Now I am ready to tell how bodies are changed
> Into different bodies.
>
> Ted Hughes, *Tales from Ovid* (1997)

Metamorphosis

Many aspects of body and behaviour obviously change, sometimes relatively suddenly, during the course of an individual's development. Sudden, discontinuous change is most obvious during the first two decades of a human life – for example, at birth and puberty. Such discontinuities are not mysteries. Many physical and biological systems are capable of changing in an abrupt, discontinuous way. Steadily increasing the pressure on a light switch does not produce a steady increase in the brightness of the bulb it controls. The switch has a point of instability, so that one moment the bulb is dark and the next moment it is fully lit. Similarly, a relatively small internal or external change can quickly transform a developing organism's characteristics to something that looks quite different. For instance, the fertilised egg of an animal rapidly divides, becoming a ball of cells, the blastula. The cells continue to divide, but do so at slightly different rates. The steady change is such that the blastula suddenly seems to collapse on one side like a deflated rubber ball and a two-layered structure called the gastrula is formed. The embryo has changed its appearance dramatically as a result of a process of continuous growth.

Sudden changes in behaviour during an individual's development may have biological utility, reflecting the changing habitat and needs of the individual as it gets older. The relatively abrupt alteration in the method of feeding at weaning, or in the mode of behaviour towards members of the opposite sex at puberty, are obvious enough. How much of the individual's personality and distinctive behavioural characteristics fails to survive the crossing of the boundary between one of the seven ages and the next?

A big change occurs in humans between the ages of approximately two and four years after birth, with the emergence of language and an awareness of self. Few adults remember much of what happened to them in their first few years. Even if they are subjectively certain that they remember their birth, say, or some early incestuous molestation, the corroboration is invariably suspect or missing.[1] It could be argued that in the first few years children have no memories; nothing has been stored so nothing has to be erased. Such a view is clearly false. Young children have good functional long-term memories. In one experiment, for example, children around two years of age were asked to imitate actions which they had seen eight months before. They performed significantly better than children who had not previously seen these actions.[2] The absence of memories from infancy does not reflect an inability to form enduring memories at the time, suggesting instead that substantial reorganisation of memory occurs early in life.

Changes in development take many forms, which invite different explanations. A qualitatively new pattern of behaviour may appear and an old one may disappear; switching from suckling to eating solid food during weaning is an obvious example. An individual child's consistent tendency to cry more than it laughs might change at a particular age. The rank-order of individuals at one age often fails to predict the order at another age, so an infant who is, say, more attentive than others at two months may be less attentive than others at four months. The slow starter sometimes catches up and overtakes the more precocious child. Parents who are proud that their child is reading at the age of four should not assume that the child will turn out a genius. Faster development does not necessarily mean a superior outcome.

Continuity from one age to the next may be lost for many reasons. One is that development is affected by many influences, not all of which are the same for everybody. Another is that children are often profoundly influenced by the social situation in which they find themselves. When Robert Hinde and his colleagues at Cambridge observed nursery school children, they found that the children's behaviour at home was not at all similar to their behaviour at school. For example, individuals who were rated by their mothers as moody engaged in few joint activities with their mothers at home, but interacted frequently with their peers in school.[3]

Continuities across age may also be lost temporarily because different children pass through a particular transition at different ages. For instance, people who are tall for their age when they are two years old are also highly likely to be tall when they are twenty. But they may not be tall for their age at thirteen, because individuals differ in the age at which they undergo the growth spurt before puberty. In this case, the property of being taller than peers survives the big changes occurring at puberty. And the same is true for many distinctive aspects of behaviour and personality. When these change permanently, as undoubtedly they sometimes do, it may not be because the person has passed through one of the metamorphoses of development.

From Elastic to Plastic

The contrasting properties of resistance to change and changeability – of elasticity and plasticity – are often found within the same material object. Stretch a metal spring a little and it will return to its former shape. Stretch it too far, however, and it will permanently take on a new shape. Adult humans, too, exhibit plasticity as well as elasticity in their behaviour, their values and their personalities; they remain recognisably the same individual in a variety of situations, yet retain the capacity to change. Compare the robustness of most people in response to life's buffetings with the way that some individuals profoundly modify their behaviour and attitudes. Continuity and change are not incompatible. The brains that generate behaviour do not consist of springs, of course, but the general property of getting back on track co-exists with an ability to alter direction.

Shakespeare understood the ability of people to change their customary behaviour and attitudes. In *Henry IV Part II*, Falstaff awaits the arrival of his young friend Hal, recently crowned as King Henry V. Falstaff calls out to Hal in his customarily familiar manner. But the king, his long-time friend, spurns him: 'I know thee not, old man.' To Falstaff's horror, Hal cold-shoulders 'the tutor and the feeder of my riots' and banishes Falstaff, on pain of death, from coming within ten miles of him. At first, Falstaff cannot believe that Hal really has changed, and insists to his cronies that the king will shortly summon him in private; the rejection, he reassures himself, is merely intended for public consumption. But it is not. Young Hal the long-time drinking companion and crony of Falstaff has now become King Henry, and in making this transformation he has put his past – Falstaff included – firmly behind him. He really has changed, by an act of deliberate will:

> Presume not that I am the thing I was;
> For God doth know, so shall the world perceive,

That I have turn'd away my former self;

Mammals, birds and other animals continue to learn and modify their behaviour throughout their lives, right up until the moment of death. Indeed, it is sometimes argued that learning helps to postpone death – at least in humans. Activating certain parts of the brain throughout life may help to maintain neuronal plasticity and thereby delay or protect the brain against the degenerative changes that accompany old age – another instance of 'use it or lose it'. The effect is clear enough in people who remain mentally active and seek out stimulation when they retire. Their mental activity may also make them physically active, so it is not always easy to be sure what really enhances their longevity. Nevertheless, as the philosopher John Dewey put it, 'The most important attitude that can be formed is that of a desire to go on learning.'

The potential to change and to carry on changing later in life is clearly important, and especially so when the effects of early experience on someone's development have been disruptive or have damaged their psychological well-being. But once someone becomes set into a particular pattern, is it really possible to wipe the slate? Distinctive features of personality and behaviour are typically formed in early life, as we discussed in Chapter 8. Sigmund Freud followed a long intellectual tradition in emphasising the formative effects of early experience. It was this view of early experience that inspired Stella Gibbons's description in *Cold Comfort Farm* of the lasting effects on old Aunt Ada Doom of seeing 'something nasty in the woodshed' when she was a child. An even more extreme view of the determination of an individual's character early in life is expressed in these opening lines from Laurence Sterne's eighteenth-century masterpiece, *The Life and Opinions of Tristram Shandy, Gentleman*:

I wish either my father or my mother, or indeed both of them,

> as they were in duty both equally bound to it, had minded what they were about when they begot me; had they duly considered how much depended upon what they were then doing; – that not only the production of a rational Being was concerned in it, but that possibly the happy formation and temperature of his body, perhaps his genius and the very cast of his mind; – and, for aught they knew to the contrary, even the fortunes of his whole house might take their turn from the humours and dispositions which were then uppermost: – Had they duly weighed and considered all this, and proceeded accordingly, – I am verily persuaded I should have made a quite different figure in the world, from that, in which the reader is likely to see me.

A reconciliation between the view that early experience is important and the view that nothing is irreversible was explicit in Freud's psychoanalytical theory. His approach, which was unusual at the time, reflected his belief that seemingly irreversible influences from childhood could be overcome in adults. This view was central to Freud's method of therapy for those whose lives had been damaged by their early experience. Nowadays the idea is widely accepted and is implicit in the vast self-help industry, which is built on the supposition that people can change themselves. Indeed, the pendulum has swung so far that it often seems as though people should be able to change their behaviour and personality as readily as they change their hairstyle.

The earlier, and perhaps excessive, emphasis on early experience may have been rejected because of its implied pessimism that once someone has missed the developmental train, nothing can be done to help them thereafter. The grounds for optimism are in fact considerable, and evidence for sensitive periods early in development may be readily reconciled with evidence for subsequent changes in behaviour. This is most clearly seen when the experience that could cause the change is not normally encountered in later life. An unwillingness to eat novel food means that people will not encounter the flavours and textures that might change their preferences. But it is not just a matter of preference. The mechanisms in the brain that protect behaviour from change were considered in Chapter 8; under rather special conditions these can be stripped away and plasticity is once again possible.

The idea that behaviour and even personality traits can change seems somehow easier to accept than the transformation of adult physical structures, built of solid flesh and bone. Someone reading the opening sentence of Franz Kafka's story 'The Metamorphosis' does not need to be told that it is a fantasy: 'When Gregor Samsa awoke one morning from uneasy dreams he found himself transformed in his bed into a gigantic insect.' And the biblical question 'Can the Ethiopian change his skin, or the leopard his spots?' was presumably intended to be rhetorical, at least in its literal interpretation. But no matter how immutable physical bodies may seem once they are fully formed, biological examples can always be found that refute the general belief in the impossibility of fundamental change. One of the most startling examples is that of certain species of reef fish.[4] They are all born female but some change sex when they are adult. The likelihood that a female will change sex depends on how many males are present in the local population. If the ratio of males to females is less than about one to ten, the socially dominant female becomes a male. The sex change depends on social status as well as sex ratio.

The behaviour patterns that are typical of gender, such as the style of play in boys and girls, may be amplified or minimised as the result of social influences from peers, teachers and parents. In the same way that a boy can become less 'boyish' in social circumstances that reduce gender differences, some shy children become less shy as they develop. Equally, some outgoing children become more withdrawn. The evidence that birth order has a significant effect on personality points once again to the subtle role of experience in development.[5] Other factors, such as sudden changes in a family's economic circumstances, can also have big effects on what happens to a child.

Breaking the Mould

Academics are sometimes caricatured (not entirely unfairly) as accumulating more and more detailed knowledge about a subject on which their focus becomes ever narrower. In the image of Waddington's epigenetic landscape, they have descended into an intellectual valley from which escape becomes increasingly difficult. But scholars manifestly do break out of these narrow confines of knowledge. Indeed, a much admired feature of the best academics is their ability to make connections between different bodies of knowledge. But is such willingness to branch out equally true for their more deeply seated beliefs and attitudes, such as their political persuasion or their sociability? Most people would say 'No'. Values are established in early life and, it is supposed, remain firmly fixed thereafter.

Political and religious leaders of various stripes throughout the centuries have sought to change people's values, and they have sometimes succeeded. How have they gone about it? A clue is given in George Orwell's portrayal of totalitarianism, *Nineteen Eighty-Four*. Winston Smith, the everyman hero of the novel, has secretly rebelled against the all-powerful Big Brother and the Party. But Winston has been uncovered, imprisoned and tortured in Room 101. His spirit is crushed and his core beliefs are re-created. Made malleable by mental and physical torture, and then exposed to the repulsive values that he has previously rejected, Winston Smith comes to love the very thing he most hated:

Winston, sitting in a blissful dream, paid no attention as his

> glass was filled up. He was not running or cheering any longer. He was back in the Ministry of Love, with everything forgiven, his soul white as snow . . . He gazed up at the enormous face. Forty years it had taken him to learn what kind of smile was hidden beneath the dark moustache. O cruel, needless misunderstanding! O stubborn, self-willed exile from his loving breast! Two gin-scented tears trickled down the sides of his nose. But it was all right, everything was all right, the struggle was finished. He had won the victory over himself. He loved Big Brother.

Comfortable in the belief that decent people are not really like that, American public opinion took a jolt in the Korean War. About a third of the 7,000 American prisoners of war collaborated with their Chinese and North Korean captors, and twenty-one refused to return to the United States when the war was over. These 'conversions' generated consternation in the United States and stimulated an intense examination of the techniques used by the Communists. Many of the apparent conversions turned out to have been little more than the consequences of severe sleep deprivation and self-preserving attempts to secure better living conditions. The so-called brainwashing methods were neither subtle nor sophisticated. Even so, some of the prisoners who had been subjected to terror, physical hardship and intensive indoctrination did seem to have changed their values and political allegiances in a more fundamental way.[6]

The origins of brainwashing lie much further back than the Korean War, however. Echoes can be found, for instance, in the Christian revivalist conversions in eighteenth-century North America. During a religious crusade in Massachusetts in the 1730s, the theologian Jonathan Edwards discovered that he could make his 'sinners' break down and submit completely to his will. He achieved this by threatening them with Hell and thereby inducing acute fear, apprehension and guilt. Edwards, like many other preachers before and after him, whipped up the emotions of his congregation to a fever pitch of anger, fear, excitement and nervous tension, before exposing them to the new ideas and beliefs he wanted them to absorb. To this day, live rattlesnakes are passed around some congregations in the southern parts of the United States; the fear and anxiety they induce can impair judgement and make the candidates for conversion more suggestible. Once this state of mental plasticity has been created, the preacher starts to replace their existing patterns of thought. And constant fear is, of course, a hallmark of totalitarian regimes, where dissenting individuals live under the unremitting threat of detention, torture or execution.

The British psychiatrist William Sargant, working in the middle of the twentieth century, was deeply interested in these mind-moulding techniques, and noticed the importance of high emotion in the process of religious conversion. In his book *Battle for the Mind*, he drew on a wide range of human experience, including that of military brainwashing.[7] He extended his inquiry to the beneficial uses of stress in psychotherapy. In the Second World War Sargant helped soldiers suffering from battle fatigue. As part of the therapy, he and his colleagues would deliberately arouse strong emotions in their patients, about events that had no direct connection with the trauma they had experienced:

> Outbursts of fear or anger thus deliberately induced and stimulated to a crescendo by the therapist, would frequently be followed by a sudden emotional collapse. The patient would fall back inert on the couch – as a result of this exhausting emotional discharge . . . but he would soon come round. It then often happened that he reported a dramatic disappearance of many nervous symptoms. If, however, little emotion had been released, and he had only had his intellectual memory of the horrible episode refreshed, little benefit could be expected.

Sargant argued for the importance of the emotional 'abreaction' in psychoanalysis, whereby patients are made anxious, guilty and even angry by their analyst and, in consequence, become able to change their previous patterns of behaviour. He quoted these comments by one of Freud's former patients:

> For the first few months I was able to feel nothing but increasing anxiety, humiliation and guilt. Nothing about my past life seemed satisfactory any more, and all my old ideas about myself seemed to be contradicted. When I got into a completely hopeless state, he (Freud) then seemed to start to restore my confidence in myself, and piece everything together in a new setting.

If Sargant was right about the benefits of the emotional experience, then that method has something in common with the psychological technique of 'flooding', in which someone suffering from a phobia is deliberately made frightened in the presence of the object or situation towards which they are phobic – for instance, by placing a large spider onto the chest of the patient who is terrified of spiders. Contrary to what intuition might suggest, the patient's phobia may be greatly reduced.[8]

It is common practice around the world for army recruits to be treated brutally in the early stages of their training. The individual is broken down through physical and mental pressure before being rebuilt in the form required by the military. The recruits are verbally abused, made to perform pointless menial tasks, forced on long marches carrying heavy equipment, and then by degrees their platoon becomes their family. Away from armies, other methods for inducing psychological plasticity include social isolation, fasting, lowering blood glucose with insulin, physical discomfort, chronic fatigue and the use of disturbing lighting and sound effects. The anthropologist Ernest

Gellner describes what is perhaps a more benign form of the same phenomenon in the creation of tribal conformism:

> The way in which you restrain people from doing a wide variety of things, not compatible with the social order of which they are members, is that you subject them to ritual. The process is simple: you make them dance around a totem pole until they are wild with excitement, and become jellies in the hysteria of collective frenzy: you enhance their emotional state by any device, by all the locally available audio-visual aids, drugs, dance, music, and so on: and once they are really high, you stamp upon their minds the type of concept or notion to which they subsequently become enslaved.[9]

The so-called Stockholm syndrome, also known as 'terror bonding' or 'trauma bonding', may be yet another instance of psychological plasticity induced by emotional trauma.[10] The term takes its name from an incident in Stockholm in the 1970s, in which a woman who was taken hostage in a Stockholm bank following an unsuccessful robbery formed a strong and long-lasting emotional bond with her captor. She even remained faithful to him during his subsequent imprisonment. Her strange reaction was not unique. Many other victims of violent hostage-taking have ended up siding with their captors against the authorities who were trying to rescue them. Being taken hostage is obviously a traumatic experience, and the hostage-takers may be equally frightened because their lives are on the line as well. In such circumstances, where hostage and captor are exposed to each other while both are emotionally highly aroused, a strong emotional bond may form, bizarrely uniting them against the world outside. As with the various military, political, religious and therapeutic techniques for changing the way adults think and behave, the crucial element is the combination of psychological stress and suggestion.

Comparable cases in which trauma has induced behavioural plasticity have been observed in other species as well. Adult wild horses are commonly 'broken' by traumatising them whilst exposing them to humans. A traditional but brutal method involves near-strangulation with a rope; even the wildest of wild horses can be reduced to gentle submissiveness in as little as fifteen minutes using this technique. Unsocialised adult dogs can similarly be induced to form strong attachments to humans by means of traumatic discipline. (We do not wish to imply that these practices are desirable simply because they work.) An anecdotal but nonetheless illuminating case concerned a remarkable change in an adult female Soay sheep, which was part of a small flock living in the grounds of the University Sub-Department of Animal Behaviour at Madingley, near Cambridge. The Soay sheep were wild, avoiding human beings, and the female in question was no exception. Then, one spring, she had a particularly difficult time giving birth. It was eventually necessary for people to assist, by catching and anaesthetising the mother and pulling her lamb out. This was undoubtedly a traumatic experience for her. Ever afterwards, until she died, this sheep remained strongly attached to humans and would follow people around as they moved about the grounds of the department. The trauma of the birth, combined with simultaneous exposure to people, brought about a profound and long-lasting change in this animal's behaviour.[11]

Stress, Love and Hormones

The concept of extreme fear or emotional arousal inducing plasticity helps make sense of many diverse examples of behavioural change. What might be the neurobiological mechanisms underlying this effect? How does trauma make someone susceptible to fundamental changes in their thoughts and values? What might be the biological link between psychological stress and the processes of plasticity and change in the nervous system?

High levels of psychological stress are associated, amongst other things, with the rapid synthesis and turnover of the neurotransmitter substance noradrenaline. This chemical messenger of the nervous system has been implicated as an enabling factor in making the adult brain become plastic again. Noradrenaline (known in the United States as norepinephrine) is released in the mammalian brain, at the endings of neurons throughout the body, and from the adrenal glands just above the kidneys. It is released, amongst other things, in response to psychological stress; in humans, a mildly stressful situation such as giving a public speech will typically elicit a 50 per cent rise in the amount of noradrenaline circulating in the bloodstream.[12]

An experiment on the visual system of cats gave some valuable insights into the connection between noradrenaline and plasticity. As we discussed in Chapter 8, the mammalian visual system is normally changeable only during an early stage in the individual's life. The capacity of an eye to stimulate neurons in the visual cortex of the cat's brain depends on whether that eye received visual input between about one month and three months after birth. If one eye is deprived of visual stimuli during this period it virtually loses its capacity to excite cortical neurons thereafter, no matter how much visual stimulation it receives. The eye consequently becomes functionally blind, even though it remains physically unimpaired. Once the dominance of the other eye is established, it is exceedingly difficult to change the relationship with the unused eye. Similarly, binocular vision cannot easily be disrupted in normally reared individuals once it has become established. However, infusing noradrenaline into one hemisphere of the visual cortex of older cats can re-establish plasticity and enable further change to occur in response to visual experience. If normally reared animals are deprived of the use of one eye during the period of noradrenaline infusion, binocular control of the neurons is lost in the visual cortex of the hemisphere that was infused. No such change occurs in the visual cortex of the other hemisphere. In other words noradrenaline can reverse in adulthood what would otherwise be unchangeable.[13]

The ability to manipulate the brain chemically, and thereby permit renewed change in adulthood, is striking. It would nevertheless be simplistic to suppose that noradrenaline alone is responsible for making neuronal connections responsive to new sensory inputs. The parts of the cat's visual cortex in which renewed plasticity occurs are also connected to neurons emanating from many other parts of the brain which may also play a role. The enabling condition for renewed change could be a particular cocktail of neurotransmitters rather than the presence of just one. Other chemical messengers besides noradrenaline are known to be involved in facilitating behavioural change. One of them is strongly associated with the state of love.

When a mother forms an emotional bond with her newborn baby the changes in her emotions and behaviour may be profound. The same can happen when someone falls in love. 'Love', wrote Stendhal, 'is like a fever which comes and goes quite independently of the will.' He characterised the onset of love as a form of crystallization:

> At the salt mines of Salzburg, they throw a leafless wintry bough into one of the abandoned workings. Two or three months later they haul it out covered with a shining deposit of crystals. The smallest twig, no bigger than a tom-tit's c[illegible] studded with a galaxy of scintillating diamonds. The [illegible] branch is no longer recognizable.
>
> What I have called crystallization is a mental process which draws from everything that happens new proofs of the perfection of the loved one.

The chemical messenger oxytocin has been strongly implicated in this process. Oxytocin has for many years been known to play a central role in stimulating the uterine contractions that occur during childbirth, and the subsequent release of milk from the mother's breast during nursing. More recently, oxytocin has been linked with the formation of strong emotional bonds between individuals who have a sexual relationship.[14] The release of oxytocin is associated with orgasm, which may not be necessary for falling in love but certainly enhances the process. It would be a mistake to pin a complex set of behavioural changes all on a single hormone. The point, though, is that the crystallisation process described by Stendhal is the result of a long cascade of events which, for a while at least, leaves the lover profoundly changed.

Something similar occurs in other species as well. Take the prairie vole, for example. Unlike most other mammals, the prairie vole is monogamous. High levels of oxytocin are associated with the initiation of the sequence of behaviour patterns which leads to the pair breeding successfully together. Oxytocin levels are lower in closely related but non-monogamous species of vole. This suggests that the hormone is especially involved in the formation of the emotional bond with a member of the opposite sex, rather than with copulation.[15] In addition to its influences on other parts of the body, oxytocin is associated with changing connectivity within the brain. The oxytocin receptors in the brain are remarkable for their plasticity. It [illegible]hat at the appropriate moments in the life cycle, when [illegible]ls must re-budget their time for the important purpose [illegible]oduction, oxytocin helps to reorganise their brains.

[illegible]xytocin is also implicated in other phenomena connected with behavioural change, including certain psychiatric disorders. The Yale University psychiatrist James Leckman and his colleagues suggested that disorders involving obsessive behaviour patterns, such as repetitive hand-washing, may have similarities to the obsessive behaviour of someone who is deeply in love. The Yale scientists found heightened levels of oxytocin in the cerebrospinal fluid of people suffering from Obsessive Compulsive Disorder. Their discoveries suggest that the sufferers' disordered behaviour results from a malfunctioning of neurobiological mechanisms which normally serve a useful role in enabling the individual's priorities to change.[16]

Long Live Change

William James, elder brother of Henry James and eminent psychologist, relished the capacity for change. In 1902 he wrote, 'The greatest discovery of my generation is that human beings can change their lives by altering attitudes of mind.' Even so, the sheer variety and complexity of behaviour and its underlying psychological systems inevitably means that any sweeping statement about the possibility of change must eventually come unstuck. The self-help industries that promise relief from shyness, depression, sloth, over-eating, or addiction to nicotine deliver results only some of the time. Once developed, some patterns of behaving are strongly buffered against subsequent change. Preferences for certain types of food and for particular places tend not to change. They may be stable for good design reasons, since change can be disruptive and costly. On the other hand, not to change may, in certain circumstances, carry an even bigger cost, which perhaps explains why behavioural characteristics tend to become plastic under conditions of stress.

When one aspect of behaviour changes it does not imply that everything else must change as well. But whatever the complexities of development and the inadequacies of current understanding, it is clear that adults really *are* capable of changing – more so, perhaps, than many suppose. As Ophelia said in *Hamlet*, 'Lord, we know what we are, but know not what we may be.'

11

Everything to Play For

> We do not receive wisdom. We must discover it ourselves after experiences which no one else can have for us and from which no one else can spare us.
>
> Marcel Proust, *A la recherche du temps perdu* (1918)

Pointless Fun

Individuals are active agents in their own development, seeking out and acquiring experiences that will change their future behaviour. Young animals and humans are equipped with developmental mechanisms that seem to have been designed specifically for this role. Collectively the behaviour is called play.

What is play? Sometimes solitary, sometimes exuberantly social, sometimes dangerous, most people know play when they see it – despite endless academic debates about its definition. Play is typically something that children and young animals do. Adults play too, of course, but generally have less time for it and less inclination. Most adults forget what it was like to spend the whole day on the beach with nothing but a bucket and spade and perhaps a friend, doing something that seems – to adults – to be entirely pointless.

Human play comes in many different forms: solitary, imaginary, symbolic, verbal, social, constructional, rough-and-tumble, manipulative, and so forth. The play of a four-year-old boy wrestling with another four-year-old is totally different from that of, say, a solitary ten-year-old staring into space whilst indulging in some private fantasy about being a pop star or a doctor. The author of *Winnie the Pooh*, A.A. Milne, painted a nice picture of solitary, imaginary play in his collection of poems, *Now We Are Six*:

> I think to myself,
> I play to myself,
> And nobody knows what I say to myself;
> Here I am in the dark alone,
>   What is it going to be?
> I can think whatever I like to think,
> I can play whatever I like to play,
> I can laugh whatever I like to laugh,
>   There's nobody here but me.

Richmal Crompton described another form of play in *Just William*, the first of many books about boys in late childhood in 1920s England. William Brown and his gang of twelve-year-old friends, 'The Outlaws', are sitting around, trying hard to think of a new game:

> They had engaged in mortal combat with one another, they had cooked strange ingredients over a smoking and reluctant flame with a fine disregard of culinary conventions, they had tracked each other over the country-side with gait and complexions intended to represent those of the aborigines of South America, they had even turned their attention to kidnapping (without any striking success), and those occupations had palled.

In the end, William comes up with an acceptable idea: 'Let's shoot things with bows an' arrows same as real outlaws used to.'

Play is quintessentially associated with having fun – and from an early age. In his observations of his first child, William Erasmus, Charles Darwin wrote:

> When 110 days old he was exceedingly amused by a pinafore being thrown over his face and then suddenly withdrawn; and so he was when I suddenly uncovered my own face and approached his. He then uttered a little noise which was an incipient laugh. Here surprise was the chief cause of the amusement, as is the case to a large extent with the wit of grown-up persons. I believe that for three or four weeks before the time when he was amused by a face being suddenly uncovered, he received a little pinch on his nose and cheeks as a good joke. I was at first surprised at humour being appreciated by an infant only a little above three months old, but we should remember how very early puppies and kittens begin to play.[1]

Tom Sawyer's Fence

While play is the antithesis of work – a welcome (and, for many adults, rare) break from the daily grind – the distinction between work and play is partly a matter of attitude. Play ceases to be play if it is imposed or taken too seriously. John Mortimer put it like this in his autobiographical work *A Voyage Round my Father*: 'At school I never minded the lessons. I just resented having to work terribly hard at playing.' Mark Twain's Tom Sawyer also sensed the gauzy boundary between work and play, and used it to his advantage. In an episode from *The Adventures of Tom Sawyer*, Tom, that 'avatar of misrule', has yet again been in a fight. Sneaking home late at night he is ambushed by his aunt, who determines to punish him. She orders Tom to spend his precious Saturday morning in hard labour: he is to whitewash her garden fence, which is thirty yards long and nine feet high. Tom is thrown into deep melancholy by the prospect, but after a few desultory brush strokes he has an inspiration: he will simply transform work into play. Another boy, Ben, appears and starts to gloat at Tom's predicament:

> 'Hello, old chap; you got to work, hey?'
>
> 'Why, it's you, Ben! I warn't noticing.'
>
> 'Say, I'm going in a swimming, I am. Don't you wish you could? But of course, you'd druther work, wouldn't you? 'Course you would!'
>
> Tom contemplated the boy a bit, and said:
>
> 'What do you call work?'
>
> 'Why, ain't that work?'
>
> Tom resumed his whitewashing, and answered carelessly:
>
> 'Well, maybe it is, and maybe it ain't. All I know is, it suits Tom Sawyer.'
>
> 'Oh, come now, you don't mean to let on that you like it?'
>
> The brush continued to move.
>
> 'Like it? Well, I don't see why I oughtn't to like it. Does a boy get a chance to whitewash a fence every day?'
>
> That put things in a new light. Ben stopped nibbling his apple. Tom swept his brush daintily back and forth – stepped back to note the effect – added a touch here and there – criticized the effect again, Ben watching every move, and getting more and more interested, more and more absorbed. Presently he said:
>
> 'Say, Tom, let me whitewash a little.'

Ben eventually persuades Tom to relinquish the whitewash brush – in return for the rest of his apple. Tom sits on a barrel in the shade, munching Ben's apple and planning the slaughter of more innocents. During the course of the day a succession of boys stop to jeer but remain to whitewash. They trade their prize possessions – a kite, a dead rat and a string to swing it with, twelve marbles, a kitten with only one eye, a brass door-knob, a tin soldier and other treasures – for the chance to indulge in the coveted task. Tom meanwhile idles the day away, accumulating wealth while the fence accumulates three coats of whitewash. Mark Twain rounds off the story by noting that 'work consists of whatever a body is obliged to do, and that play consists of whatever a body is not obliged to do. There are wealthy gentlemen in England who drive four-horse passenger-coaches twenty or thirty miles on a daily line, in the summer, because the privilege costs them considerable money; but if they were offered wages for the service that would turn it into work, then they would resign.'

Tom Sawyer exploited a characteristic of play that has since been demonstrated experimentally by psychologists – namely, that activities are more likely to be perceived as play (and therefore attractive) rather than work (and therefore unattractive) if they are entered into voluntarily. In one experiment, volunteers were given a problem-solving game to perform. Some were paid to perform the game and some were not. Those who were paid spent less of their free time performing than those for whom the only motivation was the intrinsic pleasure of the game itself.[2] Motivation to play springs from within, and the readiness to perform tasks may, paradoxically, be reduced by external rewards. A person's eagerness to play increases if the task is freely chosen and the performer discovers that their skill at some challenging task improves with practice. Success in performing the task leads to greater enjoyment and hence greater motivation to carry on. Such is the mainspring of many sports and hobbies.

Play in its many forms is, of course, not unique to humans. It is widespread among young animals of other species and may occupy a substantial proportion of their time as well.[3] At the stage in their lives when they do it most, play can account for around 10 per cent of a young animal's time – not as much, perhaps, as a child's, but still a lot. Play is observed in most mammals, some bird species, such as parrots and ravens, and is probably much more widespread in other groups than is commonly believed.[4]

Play has many attributes. In social play, for instance, the roles of the play partners are frequently reversed and sexual components are often incorporated long before the animal is sexually mature. During play involving running, jumping and other rapid movements, the movement patterns tend to be exaggerated in form, jumbled in sequence and often repeated. In some species, specific social signals are used to denote that what follows is playful rather than serious. Dogs, for example, signal their readiness to play by dropping down on their forelegs and wagging their tails, while chimpanzees have a special 'play face' which precedes a bout of social play. In the solitary manipulations of object play, the prey-catching or food-getting repertoires of adults are frequently used long before they bring in any real food. Play is also exquisitely sensitive to prevailing conditions, and is usually the first thing to go when all is not well. It is a sensitive barometer of the individual's psychological and physical well-being. For instance, young vervet monkeys in East Africa do not play in dry years, when food is scarce.[5] Play happens only when basic short-term needs have been satisfied and the individual is relaxed. It is therefore the first activity to disappear if the individual is stressed, anxious, hungry or ill.

Above all else, though, play is characterised by its apparent lack of serious purpose or immediate goal. Play is, Tom Sawyer-style, the antithesis of adult 'work', in which the behaviour has an obvious, and usually short-term, goal. Playful behaviour often resembles 'real' behaviour, but lacks its normal biological consequences: the young animal plays at fighting or catching imaginary prey, but it is usually obvious that the animal *is* playing rather than merely being incompetent.

Anyone watching young animals, such as kittens, playing is

struck by the apparent pointlessness of their behaviour. One kitten arches its back and skitters sideways at another kitten, which also arches. The first kitten leaps forward and the other simultaneously leaps into the air; it rolls over and runs away, closely pursued by the other. A few seconds later the previous attacker is attacked. The kittens end up wrestling on the floor, raking each other with their back legs. They bite each other but their jaws do not close hard enough to do any damage. Seemingly aggressive acts rarely elicit squeaks or howls as they would when adults fight. The whole thing seems utterly purposeless to the human observer. So, what is it all for?

Play has real biological costs. Animals expend more energy and expose themselves to greater risks of injury and predation when they are playing than when they are resting. Play makes them more conspicuous and less vigilant. For example, young Southern fur seals are much more likely to be killed by sea lions when they are playing than at other times.[6] The costs of play must presumably be outweighed by its benefits, otherwise animals which played would be at a disadvantage compared with those that did not and play behaviour would not have evolved. Why do young animals and children play?

From Eton to Waterloo

After his famous victory against Napoleon in 1815, the Duke of Wellington is reputed to have said that the Battle of Waterloo was won on the playing fields of Eton. The idea embedded in Wellington's probably apocryphal remark was that attitudes, discipline and team spirit acquired through playing games in childhood had brought important benefits later in life, in the 'real' world of the battlefield. The rigours of life in a British school had supposedly imbued Wellington's officers with skills that helped them prevail on the battlefield years later. (More than a century later, George Orwell, who had been at Eton himself, remarked wryly that whether or not the Battle of Waterloo was won on the playing fields of Eton, the opening battles of all subsequent British wars had been lost there.)

The belief that children's play has a serious purpose – that of acquiring skills and experience needed in adulthood – has been a central feature in thinking about the nature of play behaviour throughout history. In the sixteenth century, for example, the French essayist Montaigne wrote that 'children at play are not playing about; their games should be seen as their most serious-minded activity'. And more than two thousand years before Wellington was born, Plato was even more explicit about the role of play in development, when he wrote this in his treatise *The Laws*:

> What I assert is that every man who is going to be good at any pursuit must practise that special pursuit from infancy, by using all the implements of his pursuit both in his play and in his work. For example, the man who is to make a good builder must play at building toy houses, and to make a good farmer he must play at tilling land; and those who are rearing them must provide each child with toy tools modelled on real ones. Besides this, they ought to have elementary instruction in all the necessary subjects – the carpenter, for instance, being taught in play the use of rule and measure, the soldier taught riding or some similar accomplishment. So, by means of their games, we should endeavour to turn the tastes and desires of the children in the direction of that object which forms their ultimate goal. First and foremost, education, we say, consists in that right nurture which most strongly draws the soul of the child when at play to a love for that pursuit of which, when he becomes a man, he must possess a perfect mastery.

Plato was making the point that playful practice when young is important for the development of adult skills. In a sense play builds adult behaviour. But likening the development of behaviour to the assembly of buildings is only partly successful as an image because half-assembled animals, unlike half-assembled buildings, have to survive and find for themselves the materials they need for further construction. Nevertheless, one building metaphor – the use of scaffolding – is helpful in understanding the nature of development. Scaffolding is required for the building process but is usually removed once the job is complete. Play is developmental scaffolding. Its biological function is to help assemble the adult and, once this job is done, it falls away.

Individuals are active participants in their own development – actively engaging with the world about them in order to acquire relevant experience. Active engagement with the environment has great benefits because the world is examined from different angles – and the world rarely looks the same from different angles. The child psychologist Jean Piaget appreciated the importance of active engagement in constructing a working knowledge of the environment.[7] So many things flow from it: the recognition of objects, understanding what leads to what, discovering that things are found when stones are turned over and the world is re-arranged, learning what you can and cannot do to other people. All these discoveries are real benefits for the individual and many of them are the results of playing.

Just as with children's play, it is commonly argued that the seemingly pointless play behaviour of young animals brings biological benefits at a later stage in their development. The precise nature of those benefits remains a matter of dispute, with little hard evidence to distinguish between the possibilities. The list of putative benefits includes the acquisition and honing of physical skills needed later in life, improving problem-solving abilities, cementing social relationships and tuning the musculature and the nervous system.[3]

A notable feature of the mammalian nervous system is the superabundance of connections between neurons at the start of development. As the individual develops, many of these connections are lost and many cells die. Those neural connections that remain active are retained and the unused ones are lost. This sculpting of the nervous system is not, as is sometimes feared, an early sign of senility, but rather reflects the steadily improving efficiency of the body's classification, command and control systems. These internal changes are reflected in behaviour. When young animals playfully practise the stereotyped movements they will use in earnest later in life, they improve the co-ordination and effectiveness of these behaviour patterns. The short dashes and jumps of young gazelle when they are playing bring benefits that may be almost immediate, as they face the threat of predation from cheetah or other carnivores intent on a quick meal, and need considerable skill when escaping.[8] The cheetah's own young also need to acquire running and jumping skills rapidly in order to evade capture by lions and hyenas.[9]

Young animals may also familiarise themselves with the topography of their local terrain as a result of playing in it. Simply knowing the locations of important physical features will not guarantee rapid, safe passage around obstacles when escaping from predators or chasing prey. They need to practise. In keeping with this hypothesis, rats in a new area will typically first explore it in a cautious manner. Gradually, the speed of movement increases until the animals are running rapidly around the area along what become established pathways. Judy Stamps, a behavioural biologist at the University of California, suggested that the seemingly playful galloping ensures that, when fast movement becomes serious, the animal will be able to negotiate, efficiently and automatically, all the obstacles that clutter its familiar environment.[10] As it does so, it will be able to monitor the positions of predators, prey or hostile members of its own species. When ornithologists place mist nets to catch birds, the resident occupants of the territories in which the nets are placed drive intruding birds into these nets without being caught themselves.

Play allows the young animal to simulate, in a relatively safe context, potentially dangerous situations that will arise in their adult lives. They learn from their mistakes, but safely. On this view, play exerts its most important developmental effects on risky adult behaviour such as fighting, mating in the face of serious competition, catching dangerous prey and avoiding becoming someone else's prey. Indeed, the behaviour patterns of fighting and prey-catching are especially obvious in the play of cats and other predators, whereas safe activities such as grooming, defecating and urinating have no playful counterparts.

If play is beneficial, then it follows that depriving the young animal of opportunities for play should have harmful effects on the outcome of its development. This is, indeed, the case. For instance, the lack of play experience shows clearly in the way the animal responds to social competition. In one experiment, young rats were reared in isolation with or without an hour of daily play-fighting experience. About a month later they were put in the cage of another rat, where they were almost invariably attacked as intruders. The defensive behaviour of the play-deprived rats was abnormal. They spent significantly more time immobile than did animals that had played earlier in their lives. Other aspects of their defensive behaviour were not affected, so the effects of play deprivation appeared to be specific.[11] It seems clear that such deprivation in early life would have adversely affected the individual's capacity to cope in a competitive world. The same argument may be mounted for play-fighting in children. Through play, they learn how to cope with aggression and violence – their own and other people's.

Playing when young is not the only way to acquire knowledge and skills, however. The individual can delay acquisition until it is an adult. But when such experience is gathered without play, the process may be more costly and difficult, even if it is not impossible. As we argued earlier, many different developmental routes may be found to the same end point. Nevertheless, play has design features which make it especially suitable for finding the best way forward. In acquiring skills, individuals are in danger of finding sub-optimal solutions to the many problems that confront them. In Chapter 9 we described the epigenetic landscape pictured by Waddington, where the development of an individual organism is represented as a ball rolling down an ever-deepening valley. Suppose the ball reaches a hollow and gets stuck. It may not have reached the end of its journey, but it cannot continue unless it is jiggled about – or unless it becomes active and jiggles itself about. In deliberately moving away from what might look like the final resting point, each individual may get somewhere that is better. Play may, therefore, fulfil an important probing role which enables the individual to escape from false end points.

When individuals make active choices, changing the physical or social conditions with which they have to cope, the results may have profoundly important consequences for their subsequent lives. Individuals are usually surrounded by a wide range of environmental conditions, from which they themselves choose particular habitats, social and sexual companions and food items. The environment does not simply set a fixed problem to which the individual has to find a fixed solution. Like beavers, which create a private lake for themselves, individuals can do much to create an environment to which they are best suited.

The upshot of all this is that play almost certainly has more than one biological function, even within a single species. Some aspects of play are probably concerned with honing the development of the nervous system and musculature. Some are concerned with an understanding of future social competitors and, if it comes to it, with the martial arts that will be needed to cope with them. Some forms of play are involved with perfecting the predatory skills needed to catch prey without being injured, and some with developing efficient movement around a familiar environment to escape from predators or outwit competitors. Play is clearly not a *necessary* way for young animals to learn to recognise members of their social group, acquire knowledge of local culture, or become accustomed to their local environment. Animals are patently able to acquire these forms of experience without playing. Nevertheless, these outcomes might still be beneficial consequences of play, if and when it does occur. They were not central to the evolution of play but, once it had evolved, any additional benefit was a bonus. The young animal is able to acquire with no extra cost information of crucial importance to it, such as recognising close kin, in the course of playing for other reasons.

Human play has undoubtedly acquired yet more complex cognitive functions during the course of its evolution, re-arranging the world in ways that ultimately help understanding. While the functions of play are heterogeneous, the overarching theme is that the experience, skills and knowledge needed for serious purposes later are actively acquired and honed through playful engagement with the environment.

We have focused up to this point on play as a developmental tool used for acquiring experience which will become valuable in a later stage. But play does not stop altogether after childhood, and adults can benefit from playfully updating their knowledge and skills. When humans perform actions at high speed, like playing a new piano part, they start out relatively slowly, monitoring each movement they make. As they get better they are able to speed up, producing long sequences which become so automated that a passage cannot easily be started in the middle. Athletes are encouraged not to think about their actions. Indeed, they often have to do things so quickly that they do not have time to think. Professional tennis players facing balls that may be served at them at 120 miles per hour or more have to make split-second decisions again and again throughout a match. Repeated play – which is doubtless seen as work by professional athletes – not only improves physical fitness, it also organises complex movements so that they can be performed quickly, accurately and automatically. In a different sphere altogether, play can also be a powerful source of innovation and creativity in adult life.

Connecting the Unconnected

Picasso was once filmed painting onto glass. The onlooker saw the picture emerge, but viewed from the other side of the glass. Picasso started by quickly sketching a goat and then rapidly embellishing it. Other shapes appeared and disappeared; colours were mixed and transformed. By the end of the film the goat had long since gone and it would have been hard to say what the picture was all about. Picasso had been playing – probably showing off – but clearly enjoying himself hugely. In a similar spirit, friends or colleagues faced with an interesting problem may toss their ideas backwards and forwards, mixing different thoughts and approaches in new combinations. On a good day, their playfulness may lead them to fruitful new pastures and the process is great fun.

Creativity and innovation are all about breaking away from established patterns. The writer William Plomer put it like this: 'It is the function of creative men to perceive the relations between thoughts, or things, or forms of expression that may seem utterly different, and to be able to combine them into some new forms – the power to connect the seemingly unconnected.' In the same vein William James described genius as 'little more than the faculty of perceiving in an unhabitual way'.

Play is an effective mechanism, therefore, for facilitating innovation and encouraging creativity. Playfully re-arranging disparate thoughts and ideas into novel combinations – most of which will turn out to be useless – is increasingly used in business and in academic research as a powerful means of gaining new insights and opening up possibilities which had not previously been recognised. Management consultants have noticed the creative benefits of playful thought, hence the established practice in business organisations of conducting brainstorming sessions. Play, in other words, extends to pure thought. It involves doing novel things without regard to whether they may be justified by a specified pay-off. Understandably, such activities, like the play of animals, are usually the first to go when times get hard. The drying up of funds for what are perceived as luxuries is the first sign of a struggling company. Organisations in trouble and short of resources typically become risk-averse and focus on short-term goals at the expense of their long-term ability to innovate. In the long run, many go to the wall because their competitors overtake them with innovative products or services.

Periodically, the government agencies which fund scientific research try to assess how successful the scientists' activities have been. They are usually disappointed. The frequency with which the scientists they have supported have hit the jackpot is embarrassingly low. Those controlling the purse strings often draw the wrong conclusion, namely that they should only fund safe, predictable research which follows a well-trodden path. They become risk-averse, particularly when money is in short supply, insisting that every step of a research programme must be worked out in advance and specified in the grant application. The obvious riposte is that if the outcome is known in advance then why bother to do the research?

Major advances in understanding simply do not occur if intellectual risks are never taken and attention is focused solely on short-term benefit. The result is mediocrity. The uncertainty involved in all artistic and scientific exploration is obvious. It is hard to predict when such activities will change the shape of human understanding. But one thing is clear. Nothing will change if playful creativity is stultified by excessive caution.

Scientists often describe their work as playful. Isaac Newton (who was described by William Cowper as that 'childlike sage') described his own scientific genius in these self-effacing terms:

> I don't know what I may seem to the world, but as to myself, I seem to have been only like a boy playing on the sea-shore and diverting myself in now and then finding a smoother pebble or a prettier shell than ordinary, whilst the great ocean of truth lay all undiscovered before me.

James Watson's book *The Double Helix* gives a personal account of how he and Francis Crick, working in Cambridge in the early 1950s, discovered the double-helical structure of DNA. This was one of the major scientific achievements of the twentieth century, for which Crick and Watson, together with Maurice Wilkins, won the 1962 Nobel Prize. *The Double Helix* conveys strongly the playful nature of scientific creativity; the way in which scientists bat ideas back and forth, trying out new combinations and discarding them if they do not pass muster.

Crick and Watson's intellectual attack on the puzzle of DNA was swift and brilliant. Many of their best ideas were hatched over enjoyable lunches in the Eagle, a pub in the centre of Cambridge. Watson and Crick seized upon model-building as a way of rapidly testing different theories of the DNA structure, an idea that the American chemist Linus Pauling had used to great effect in working out the alpha-helical structure of protein. The essential trick had been to ask which atoms fit next to each other. Their main working tool had been a set of coloured balls superficially resembling the toys of pre-school children. Watson wrote, 'All we had to do was to construct a set of molecular models and begin to play – with luck, the structure would be a helix.' And, of course, it was. As everything fell into place, Watson and Crick felt, not without reason, that they had discovered the secret of life. The morning after their discovery Watson walked through the middle of Cambridge, staring up at the Gothic pinnacles of King's College chapel as they stood out sharply against the spring sky. He reflected on how much of their success had been due to the long uneventful periods when he and Crick had simply walked through the beautiful courts and gardens of the colleges.

Roaming Freely

If theories about the importance of play in human development are correct, then the tendency of some parents to over-protect their offspring from all sorts of real and imagined dangers has worrying implications for their children's development. In Britain and the United States, parental concern for children's security has led to profound changes in the nature of childhood. Once-normal activities such as roaming freely about with friends, or even simply walking unescorted to and from school, are becoming increasingly rare experiences. It is widely believed that children should not be left on their own, and the activities of many children are constantly monitored to ensure that they do not come to any harm. In his discussion of the growing but largely groundless anxieties of parents in Western societies, the British sociologist Frank Furedi commented, 'It is in the sphere of children's lives that the institutionalisation of caution has had the most far-reaching effect. During the past twenty years, concern with the safety of children has become a constant subject of discussion. Children are portrayed as permanently at risk from danger.' But despite impressions to the contrary, the incidence of child molestation and murder has remained relatively constant in the majority of developed counties in peacetime.[12]

Children are less and less able to muck about. They are increasingly deprived of opportunities to play-fight and to socialise, and hence may find it harder to cope with conflict or to co-operate. They take less and less physical exercise, becoming fat and unfit. Schools increasingly regard play as unproductive 'down-time'. Scheduled playtime (or 'recess' as it is known in the United States) is squeezed out by the pressures of an expanding formal curriculum; this is also done, of course, with an eye to competitive league tables and the demands of ambitious parents. In Chapter 3, we discussed the long-term benefits of relatively unstructured pre-school activity, as revealed by the High/Scope project. The socialising role and the part that such activities have in motivating children in their subsequent education have profound implications for how childhood play should be regarded. Childhood is a time for experiences that may never recur.

12

Sex, Beauty and Incest

> There is no excellent beauty that hath not some strangeness in the proportion.
>
> Sir Francis Bacon, 'Of Beauty', *Essays* (1625)

It's Just It

When somebody makes a conscious choice of a sexual partner, what aspects of their development and early experience influenced that choice? The answer is likely to depend on what that person had in mind – a short-term fling or a lifelong commitment. Recognising this distinction, John Donne wrote, 'Love built on beauty, soon as beauty, dies.' As for the source of attraction, Rudyard Kipling suggested, 'T'isn't beauty, so to speak, nor good talk necessarily. It's just It. Some women'll stay in a man's memory if they once walked down a street.'

Discussion of the origins of sexual attraction, perhaps more than any other topic in human biology, triggers a nature–nurture debate that becomes especially intense when it turns to the sexual preference that some people have for partners of their own sex. We hope that a cool look at what is known about development can lead to some understanding of the processes involved and also to greater sensitivity towards the behaviour of others. The links between the biology of mate choice and the anthropology of marriage and incest taboos are another fraught area. Once again, the touchstone for understanding comes from knowing how behaviour develops. We shall look first at how sexual partners are chosen and how these preferences develop.

Reflection on what draws people to each other is a recurring theme in literature. Here is the evocative moment in Tolstoy's *Anna Karenina* when Count Vronsky meets Anna on a train and immediately falls in love with her:

> Vronsky followed the guard to the carriage, and at the door of the compartment had to stop and make way for a lady who was getting out. His experience as a man of the world told him at a glance that she belonged to the best society. He begged her pardon and was about to enter the carriage but felt he must have another look at her – not because of her beauty, not on account of the elegance and unassuming grace of her whole figure, but because of something tender and caressing in her lovely face as she passed him. As he looked round, she too turned her head. Her brilliant grey eyes, shadowed by thick lashes, gave him a friendly, attentive look, as though she were recognizing him, and then turned to the approaching crowd as if in search of someone.

In a major study organised by the evolutionary psychologist David Buss, the desirable characteristics of individuals of the opposite sex were investigated in more than 10,000 men and women from thirty-seven cultures around the world. Buss and his colleagues found that both men and women rated the capacity for love, dependability and emotional stability highly. But big differences were found between men and women in what characteristics were valued in a member of the opposite sex. Women rated especially the resources held by a man and all the characteristics associated with acquiring such resources, such as health and intelligence. Men rated youth and physical attractiveness more highly than did women. These sex differences were consistent across all cultures.[1] In a separate study of the preferences of both Caucasians and Japanese, the most attractive female face had larger eyes, higher cheekbones and a thinner jaw than the average. This face also had shorter distances between nose and mouth, and between mouth and chin.[2] Despite great variation in preferences for plumpness between cultures and across time within cultures, men typically prefer women whose waist-to-hip ratio is 0.7.[3]

Jane Austen famously opens *Pride and Prejudice* as follows: 'It is a truth universally acknowledged, that a single man in possession of a good fortune, must be in want of a wife.' And it is not only wealth that matters. Social class, race, intelligence and health all play their part. One study found that attaching national labels to people affected how physically attractive they seemed to the opposite sex. People of differing nationalities, perceived as varying in status, were rated on their physical attractiveness. The assessments of physical attractiveness were found to be correlated with the ranking for national status.[4] In other words, individuals from high-status nations were on average regarded as more physically attractive. The effect also worked in the reverse direction, so that people who were judged to be attractive without knowledge of their nationality were assumed to be from higher-status nations. These prosaic and static details do not capture the importance of the responsiveness so keenly appreciated by Tolstoy when he described Vronsky's attraction to Anna Karenina. The sense that the other person is interested means a lot. So, indeed, do many other attributes – particularly when looking for a long-term partner.

Extravagant Ornaments

Most men and women think carefully about their appearance when attempting to attract the opposite sex, hence the obsessions with fashion, ornamentation, slimness and make-up. Decorating the body in order to interest others is as old as recorded human history. The link to what is found in animals is tantalising, since an important aspect of mate choice is the role of particular physical characteristics. Biological inheritance will ensure that these will be passed on to offspring and attract mates to those offspring when they become adults. The female peacock, for example, prefers to mate with males who have on their tails more spots that resemble eyes.[5] If her sons have more of these eye-spots on their tails than offspring fathered by a male with fewer spots, those sons will themselves attract more mates, and the female will consequently end up with more grandchildren. This aspect of mate choice, which has remarkable evolutionary consequences, was first identified by Charles Darwin in another of his important books.[6] The evolutionary principle is known as 'sexual selection'. Initially, Darwin's ideas about this aspect of evolution were ridiculed or ignored, but a century later the role of ornament in mate choice has become one of the major areas of research in animal behaviour.

The possibilities for simple experiments to test theories about the importance of animal ornaments are endless. For instance, the long-tailed widow bird of East Africa is about the size of a starling, but during the breeding season the male grows a tail about 50 cm long. The tails of males may be artificially shortened or extended by gluing in extra pieces of tail. When this was done, significantly more females were attracted to the males with extended tails.[7] The long tail is evidently important when the female chooses a mate. But the size to which a male's tail can usefully grow is limited; in wet weather the long tail becomes drenched with water and the male may then be unable to escape from predators on the ground. A balance must therefore have been struck during the course of evolution between survival and attracting mates.

Biologists debate why such seemingly bizarre devices for

attracting mates should have evolved. Some regard them as an inevitable chance consequence of mate choice: once enough individuals have them, the theory goes, then both the preference and the chosen characteristic will drive each other in a runaway evolutionary process.[7] Other biologists believe that the characteristic chosen by the opposite sex is an indicator of true biological worth – the bigger the tail, for instance, the stronger must be the animal which can produce and carry it.[8] Another possibility is that the evolutionary process depended on the combination of two mechanisms: preferring a mate who is slightly unfamiliar and, in birds, having feature-detectors that respond to eyes and, hence, eye-spots on tails. So a peahen prefers a peacock with slightly more eye-spots on his tail than familiar males have on their tails. The average number of eye-spots on tails ratchets up in each generation, since the males that have more eye-spots collect most mates.[9] Whatever the evolutionary explanation, and those mentioned are not mutually exclusive, the significance of the phenomenon identified by Darwin is no longer disputed.

The obvious importance of physical features in attracting members of the opposite sex has led to much speculation about whether particular non-physical features of humans may also be an evolutionary product of mate choice. Is it the case, say, that large vocabularies, musical and visual art, or risk-taking behaviour serve the function of attracting the opposite sex?[10] If so, the argument runs, once they had become important in human evolutionary history, these behavioural ornaments might have ratcheted upwards, just like the eye-spots on the peacock's tail, because individuals with bigger vocabularies or better artistic skills would have had more descendants. Biology is rarely that simple, but the conjecture does raise questions some of which can be answered. Risk-taking behaviour, such as driving too fast, is certainly characteristic of young adult human males. But is it really the case that humans produce their most extravagant vocabulary, music and art at an age when they do most of their courtship, and that they do so in the presence of one or more members of the opposite sex?

Some Strangeness in the Proportion

What makes a face beautiful? Francis Galton, Charles Darwin's highly inventive cousin, was the first to experiment with the technique of averaging photographic images of human faces. He superimposed images of different faces, adjusting the projected image of each negative so that the positions of the eyes, noses and mouths of all the faces corresponded on the final picture, which was produced by multiple exposures. Even with the relatively crude photographic methods of the nineteenth century, Galton was able to obtain remarkably sharp pictures of average faces.[11] With modern technology, photographs can be digitised and the images manipulated on a computer, so that many individual faces are superimposed to produce an average. It is possible to portray an average twenty-year-old, an average forty-year-old, an average Englishman, an average Greek woman, an average Japanese child, and so on. The faces are surprisingly beautiful and have prompted speculation that similar averaging processes happen in the course of everyday human experience, setting aesthetic standards that have real biological utility.

Average human faces are highly symmetrical, and links have been drawn between this fact and the behaviour of other animals when given choices between symmetrical and asymmetrical members of the opposite sex. Females of several species prefer to mate with males who have more symmetrical sexual ornaments. For instance, the common barn swallow has long tail streamers; the closer these are to each other in size, the more strongly is the male preferred by a female.[12] Bilateral symmetry may be important in mate choice because it indicates the individual's true ability to cope with ill health and environmental stress. Efficient regulation of development, it is argued, leads to greater symmetry. Symmetry may therefore provide a reliable indicator of mate quality. Some evidence suggests that humans of both sexes prefer individuals of the opposite sex who are symmetrical in their face, body shape and, in the case of women, their breast size.[13]

And yet average faces with all their irregularities ironed out look bland. They do not seem as attractive as less regular faces. Some degree of asymmetry or quirkiness can make a face more appealing. Louis de Bernières makes this point in his novel *Captain Corelli's Mandolin*, set on the Greek island of Cephalonia in the Second World War. Deepening love develops between Pelagia, a bright-spirited local girl, and Captain Corelli, an officer of the occupying Italian army. When it comes to beauty, Corelli regards symmetry as only a property of dead things:

> It's fine for buildings, but if you ever see a symmetrical human face, you will have the impression that you ought to think it beautiful, but that in fact you find it cold. The human heart likes a little disorder in its geometry, Kyria Pelagia. Look at your face in a mirror, Signorina, and you will see that one eyebrow is a little higher than the other, that the set of the lid of your left eye is such that the eye is a fraction more open than the other. It is these things that make you both attractive and beautiful, whereas . . . otherwise you would be a statue.

Some of the confusion surrounding the relationship between beauty and attractiveness has been cleared away by the British physiologist David Perrett, who has extended his study of the face-detectors in the brain to an analysis of how people react to images of human faces.[2] He and his colleagues found that faces are perceived as more attractive if some of the facial features are exaggerated by caricaturing the image so that it differs from the average. This work and other studies have shown that people are most attracted to faces that are distinctive. Regularity on its own is not enough; average faces are perceived as beautiful, but attractive faces depart slightly from the average.

A preference for faces that are a bit different, but not too different, from a familiar standard is relevant to mate choice in other species. Animals of many species tend to avoid mating with individuals who are very close kin, such as siblings, but they do sometimes prefer to mate with more distant relatives. Japanese quail, for example, prefer mates who are first or second cousins, when given a choice in laboratory experiments. If they have been reared with unrelated individuals, the quail prefer mates that are a bit different from these familiar individuals. Humans choose partners somewhat like themselves. At the same time, people prefer sexual partners who look slightly different from individuals with whom they have grown up.[14]

Natural experiments, somewhat similar to the laboratory studies with quail, have been performed unwittingly on human beings. Famously, Israeli kibbutzniks grow up together like siblings and rarely marry each other.[15] The most comprehensive evidence has come from marriage statistics from Taiwan in the nineteenth and early twentieth centuries, when Taiwan was under Japanese control. The Japanese kept detailed records for the births, marriages and deaths of everyone on the island. As in many other parts of south-east Asia, marriages were arranged, and occurred mainly and most interestingly in two forms. The major type of marriage was the conventional one in which the partners first met each other when adolescent. In the minor type of marriage, the wife-to-be was adopted as a young girl into the family of her future husband. In minor-type marriages, therefore, the partners grew up together like siblings. In this sense they were like the quail in the laboratory experiment, having been reared with an individual of the opposite sex to whom they were not genetically related. Later in life their sexual interest in their partner was assessed in terms of divorce, marital fidelity and the number of children produced. By all these measures, the minor marriages were conspicuously less successful than the major marriages. Typically, the young couples who had grown up together from an early age, like brother and sister, were not much interested in each other sexually when the time came for their marriage to be consummated.[16]

Striking the Balance

Humans, like many other birds and mammals, usually grow up together with their close kin. Preferring a member of the opposite sex who is a bit different from close kin will minimise the biological ill-effects of inbreeding. If an animal inbreeds too much, it might as well multiply itself without the effort and trouble of courtship and mating. Conversely, a preference for an individual somewhat like close kin will minimise the opposing ill-effects of breeding with individuals who are genetically too different. A sexual preference for individuals who are a bit different from close kin strikes a balance between the biological costs of inbreeding and those of outbreeding.[14]

The biological costs of inbreeding are evident enough, particularly in birds. If a male bird is mated with his sister and their offspring are mated together, and so on for several generations, the line of descendants usually dies out fairly quickly. This happens because of the accumulation of recessive genes. Some potentially harmful genes are recessive and therefore harmless when they are paired with a dissimilar gene, but they become damaging in their effects when combined with an identical gene. They are more likely to be paired with an identical recessive gene as a result of inbreeding. The presence of such genes is a consequence of the mobility of the birds and the low probability that they will mate with a bird of the opposite sex that is genetically similar to them. Over time, the recessive genes have accumulated in the genome because they are normally suppressed by their dominant partner gene. The genetic costs of inbreeding arising from the expression of damaging recessive genes are the ones that people usually worry about. However, recessive genes are less important in mammals than they are in birds because mammals generally move around less and may live in quite highly inbred groups. The most important biological cost of excessive inbreeding is that it negates the benefits of the genetic variation generated by sexual reproduction.

On the other side, excessive outbreeding also has costs. For a start excessive outbreeding disrupts the relation between parts of the body that need to be well adapted to each other. The point is illustrated by human teeth and jaws. The size and shape of teeth are strongly inherited characteristics.[17] So too are jaw size and shape, as may be seen in the famous paintings of the Hapsburg family scattered round the museums of the world. The Dürer painting of the Holy Roman Emperor Maximilian I reveals the large Hapsburg jaw, which remained as pronounced in his great-great-great-grandson, Philip IV of Spain, shown in the portrait by Velázquez. The potential problem arising from too much outbreeding is that the inheritance of teeth and jaw sizes are not correlated. A woman with small jaws and small teeth who had a child by a man with big jaws and big teeth lays down trouble for her grandchildren, some of whom may inherit small jaws and big teeth. In a world without dentists, ill-fitting teeth were probably a serious cause of mortality. This example of mismatching, which is one of many that may arise in the complex integration of the body, simply illustrates the more general cost of outbreeding too much.

Close relatives, such as fathers and daughters or brothers and sisters, who were separated before experience inhibited their mutual sexual interest, have sometimes been reported as finding each other extremely attractive. Precisely what cues they respond to in each other is far from clear, but the phenomenon seems real. Normally such a process would combine with the effects of experience to generate a preference for somebody slightly different from a close relative. The evolutionary pressures that generated these and other psychological mechanisms arose from too much inbreeding, on the one hand, and from too much outbreeding on the other. The result was a balance in which individuals had greatest reproductive success if they mated with a partner who was somewhat similar to themselves, but not too similar.

The face is self-evidently central to sexual attractiveness in humans. But smell also plays an important role in human sexual attraction, as it does in other species. In rats and mice, mate choice is heavily influenced by odours that carry cues about the degree of genetic relatedness. Once again, ideal mates are genetically dissimilar but not too dissimilar. Inbred mice prefer to mate with individuals that differ from them only in a few genes in the Major Histocompatibility Complex (MHC). The proteins coded for by MHC genes act as markers on cell surfaces and play an important role in controlling immune responses. Even individuals who are genetically almost identical still differ in some MHC genes. These genes therefore provide a good indicator of genetic relatedness. Breakdown products from the MHC proteins are found in the urine and appear to affect its smell. Mice can detect these subtle differences in smell and thereby the degree of genetic relatedness to themselves.[18]

Some traditional customs imply a role for smell in humans. For example, in the Philippines the ceremony marking engagement ends with the couple exchanging items of clothing, which they kiss and smell. In an experiment where men and women rated the pleasantness of smell of T-shirts worn by other people, the other person's body odour was perceived to be more pleasant if that person was genetically more dissimilar. In humans the genes of the MHC may influence an individual's body odour.[19] Here again, sexual preferences for slight differences from self or close kin would help to secure optimal outbreeding.[20]

Cooking the Preferences

Debates about the origins of beauty and attractiveness readily degenerate into the familiar but unproductive split between nature and nurture. Claims about biological universals are countered by claims about the cultural relativity of aesthetic judgements, which shift from one generation to the next. Theories about the biological utility of sexual preferences are countered by charges of story-telling and by evidence that culture influences those preferences. These oppositions are often false. The preferred fatness or slimness of the ideal woman may be culture-bound, while the ideal ratio of waist to hip is not. If a male preference for a waist-to-hip ratio of 0·7 is, indeed, universal, the *development* of such a preference still needs to be explained. Whatever that explanation turns out to be, the average male preference for female body shape is likely to be the product of at least two separate developmental processes, one of which may be universal and the other of which clearly is not. Furthermore, the claim about universality should not be accepted uncritically – even from a Darwinian perspective.

Undoubtedly, the cross-cultural similarities found by David Buss and his colleagues punctured the belief that particular mating preferences are strongly tied to particular cultures.[1] However, the attempts by evolutionary psychologists to establish universals, thereby debunking the relativism of what they refer to as the Standard Social Science Model,[21] may have missed a point – namely that for sound evolutionary reasons, humans are adapted to their local environments. In Chapter 6 we suggested that humans might be so adapted. Men are much closer in size to women in harsh environments, in which monogamy is the best option.[22] Similarly, the combination of characteristics used in choosing a partner may reflect the ways in which men resemble women in their commitment to caring for one family in such conditions.

Babies gaze for longer at faces that are judged by adults to be attractive.[23] This may be because their face-detectors are tuned before birth to particular features and these detectors remain in place throughout their lives. On the other hand, human babies are perceptually competent at an extremely early age and quickly form prototypes from the sets of faces which they have seen from immediately after birth.[24] These would be beautiful averages. It seems likely that the development of some feature-detectors is not dependent on learning, while the development of others does depend on the specific effects of experience. Evidence that people prefer members of the opposite sex who are a bit different from those they have grown up with certainly suggests that recognition of individual faces is important. The sexual preferences of many birds are strongly influenced by their early experience. Indeed, if they are exposed to a different species of roughly comparable appearance during the sensitive period, they prefer to mate with a member of that host species rather than a member of their own species. We have already considered the much more subtle effect which enables birds to recognise the individuals with whom they grew up and choose a slightly different partner. Sexual imprinting, as it is called, is not complete until the birds have developed their distinctive adult plumage. In that way it is different from filial imprinting (learning about parents) which is complete at a much earlier stage in the life cycle. The timing of the developmental process is nicely linked to the ecology of the species.[14]

Early experience with individuals of the opposite sex seems important in humans and the age when that experience starts is crucial. The few Israeli kibbutzniks who chose to marry within their peer group were usually those who had entered the kibbutz after the age of six and therefore had not grown up with their future spouses.[15] In Taiwan, girls who were adopted into families before the age of three and then married their adopted 'brother' had a lower fertility than girls adopted later.[16] Neither of these findings means that the learning process which affects adult sexual preferences is completed early in life. If children grow up together and, as a result, see a lot of each other, they revise the ways in which they recognise each other; this goes on until they are sexually mature. By the time they are three, children are highly conscious of their own sex and are much less likely to play with somebody of the opposite sex, particularly a child who is not well known to them.[25] It seems plausible then that a girl who is adopted when she is over the age of three will be viewed as a stranger by the boy, and treated differently from a girl who is adopted when younger.

Like faces, individual smells may also be learned and lead to preferences for mates that are a bit different. Here again, smell forms the basis for a standard and then something slightly different is preferred in adult life. In both facial and smell recognition, genetic differences are linked to the facial or odour differences between people. When something is known about the development of sexual preferences, the nature–nurture dichotomy becomes less appealing.

The Developmental Origins of Homosexuality

Some humans prefer their own sex or have enjoyable sexual experiences with both men and women. The phenomenon is not recent. The poet Sappho lived on the Greek island of Lesbos more than 2,500 years ago and referred explicitly in her poems to her love for other women. Here is a translation of one of the remaining fragments:

> For I only, briefly, need glance at you to find my voice has gone and my tongue is broken, and a flame has stolen beneath my skin, my eyes can no longer see, my ears are ringing, while drops of sweat run down my trembling body, and I've turned

paler than a wisp of straw and it seems to me I'm not far off dying.

How is it that some people prefer their own sex? Many attempts have been made to provide a simple answer. In Chapter 9 we described the capacity of human behaviour to change in particular states, such as that of love. Once in a particular valley of the developmental landscape, the willingness and the means to switch to another is reduced. So the linkage between falling in love for the first time and a homoerotic experience might seem a plausible explanation for the homosexual preferences of some people. But attempts to pin homosexual preferences on a particular type of formative experience have not been successful.

Given an understandable quest for simplicity, it was perhaps inevitable that a putative gene 'for' homosexuality would also infect the discussion of its development. A 'gay gene' was duly identified. However, the gene was shown to be non-specific in its influences, affecting many aspects of behaviour.[26] One study of twins suggested a genetic basis for male homosexuality; of those who had a homosexual identical twin, 52 per cent were also homosexual, whereas of those who had a homosexual non-identical twin only 22 per cent were homosexual. However, the same study also found that only 9 per cent of non-twin brothers of homosexuals were also homosexual, almost the same proportion (11 per cent) as for adopted brothers. If homosexuality had been inherited on a simple genetic basis, then the proportion of non-twin brothers who were homosexual should have been higher than for genetically unrelated adopted brothers and no different from non-identical twin brothers.[27] The results indicate that non-genetic factors also play a major role in the development of homosexuality in males.

Other evidence links male homosexuality with the presence in the family of older brothers. Males who have older brothers are statistically more likely to be feminine homosexuals. (By no means all male homosexuals are feminine.) The greater the number of older brothers, the greater the likelihood of the boy being homosexual. No association is found with the number of older sisters, younger siblings or parental age. Even when they are young, boys with older brothers are more likely to behave in feminine ways.[28] A possible mechanism which would explain these findings is that during pregnancy mothers are progressively immunised by proteins derived from the Y chromosome and only produced by males. These proteins might pass through the placenta and lead to the formation of antibodies in the mother. When the mother is pregnant with another boy her antibodies might pass through the placenta and attack the male-linked proteins; as a result, it is suggested, the subsequent sons are less masculine.[29] Other evidence also points to subtle influences of the uterine environment on physical development. Homosexual men are more likely than heterosexual men to have leftward asymmetry in the dermal ridges on their fingertips.[30] These ridges are formed early in pre-natal development and do not subsequently change. Fingerprint development is affected by maternal blood pressure, but just how that is linked to the development of sexual orientation is far from clear.

The structure of the brain is influenced by exposure to various sex hormones and neurotransmitters during sensitive periods in early development.[31] Under conditions of stress, or as a result of genetic mutations, pregnant mothers may produce male hormones (androgens) from their adrenal glands. These hormones influence brain development in the foetus. Exposure to high levels of androgens during a sensitive period for brain development produces a predisposition towards homosexuality in females and heterosexuality in males.[32] Low levels of androgens have the opposite effect. Environmental influences on sexual development continue after birth. Later in life the brains of male homosexuals are found to differ structurally in certain respects from those of heterosexuals – notably in the hypothalamus, which is involved in the regulation of sexual behaviour.[33] Sexual differentiation of the human brain continues for at least two to four years after birth, which means that it may be influenced by other environmental factors, such as the release into the environment of man-made substances that mimic the action of female hormones.

As in so much else, though, the various chains of events from sundry predisposing factors to adult behaviour are long and indirect. A particular gene or hormone might influence the brain and thence behaviour. A slight tendency to behave in a way atypical of others of the same sex may lead children to choose circumstances in which they feel more comfortable ('niche-picking'). These conditions, more typical of the opposite sex, feed back onto the way the individuals' brains are developing. This then amplifies their styles of behaviour and so on. Childhood behaviour affects the character of experience which, in turn, helps to shape sexuality. Such plausible but complex cooking of behaviour should not simplified into an either/or, genes/environment formulation.

Scientific theories about the developmental origins of homosexuality are undoubtedly laden with social and political significance and generate contradictory responses. Some see the theories as oppressive, leading to attempts at 'cures' and suppression of the way that homosexuals have chosen to live their lives. Others see the biological theories as liberating, leading to greater acceptance of their sexuality and their preferences. As things stand, the scientific evidence is complex and equivocal. It does, however, show that the cascade of events leading to homosexuality sometimes starts early in life in both sexes. Many explanations exist alongside each other and, in the case of any one developmental process, the interplay between the developing individual and the environment through which he or she passes is crucial.

Serious Commitment

A preference for an individual of the opposite sex usually leads sooner or later to mating and eventually to the production of offspring. Many marine animals simply release their sperm and eggs into the environment in vast numbers; fertilisation occurs remotely from the parents and parental investment is limited to sperm and egg production. The species that take much time and trouble over finding the right mate and caring for their young are primarily birds and mammals, but many examples of extensive parental care are found in the cold-blooded vertebrates, and exquisite examples are found in the insects and crustacea. In birds, both sexes commonly share the burden of looking after their offspring. In many species of fish, the male cares for the eggs once he has fertilised them. Paternal care in mammals is relatively rare in mammals, but some fathers do play an important role. For example, the California mouse is monogamous and has much greater success with litters of four pups if both parents look after the young.[34]

Mating arrangements within a particular species can be surprisingly flexible, as in the case of the dunnock, a drab little bird that was once regarded as a zoological model for Victorian family values. This is how a nineteenth-century clergyman naturalist, the Reverend F.O. Morris, described it:

> Unobtrusive, quiet, and retiring, without being shy, humble and homely in its deportment and habits, sober and unpretending in its dress, while still neat and graceful, the Dunnock exhibits a pattern which many of a higher grade might imitate, with advantage to themselves and benefit to others through an improved example.[35]

Despite her reputation for upholding monogamy, the female dunnock may in fact mate with more than one male when her first mate is not looking. If she is successful, all the fathers of her offspring participate in caring for the chicks after they have hatched. However, in territories that are particularly well supplied with food, a male may mate with more than one female.[36] The behavioural ecologist Nick Davies, who made these discoveries in the Botanic Gardens at Cambridge, also found more complex mating arrangements, in which, for example, a female shared a principal mate with another female but also received help from another male in caring for her young. Both the male and the female dunnock are opportunists, each maximising its own reproductive success in ways that are tailored to prevailing conditions.

As with the dunnock, enormous variety is found in human child-rearing practices. Monogamous marriages are required in only about a third of human societies. In polygynous societies, in which a man may have more than one wife at a time, the sharing of parenting is unequal but the man is still expected to provide the resources to enable his many wives to care for his offspring. In the few polyandrous societies, in which the woman may have more than one husband at a time, the husbands look after their wife and her children. Humans might appear to be like dunnocks in having flexible offspring-rearing arrangements which reflect local ecological conditions. The analogy would, however, obscure an important difference. In many human societies, monogamous as well as polygynous, the choice of partner is not made by the people concerned. Rather, the decision is made by their parents, in consultation with others who are supposedly knowledgeable about such matters. The structure of purely human institutions sometimes bears little relationship to individual desire.

The survey by David Buss and his colleagues of thirty-seven different cultures found that the importance of love in a sexual relationship was valued more highly than anything else by both sexes. Clearly the experience of romantic love is not restricted to the cultures in which partners are freely chosen or those in which monogamy is the custom. Irrespective of local marriage rules, men and women entering a relationship are commonly drawn together by physical appearance. Here, as she fell in love, is the end of Molly Bloom's monologue in James Joyce's *Ulysses*:

> Gibraltar as a girl where I was a Flower of the mountain yes when I put the rose in my hair like the Andalusian girls used or shall I wear a red yes and how he kissed me under the Moorish wall and I thought well as well him as another and then I asked him with my eyes to ask again yes and then he asked me would I yes to say yes my mountain flower and first I put my arms around him yes and drew him down to me so he could feel my breasts all perfume yes and his heart was going like mad and yes I said yes I will Yes.

During the 'honeymoon' period, incompatibilities of behaviour are ignored. It is only later, when the effects of physical attraction have started to wane, that a mismatch of styles and values might disrupt the relationship. But in the intervening period, one or both partners may have changed their own ways of behaving so that they mesh more satisfactorily with each other. In Chapter 10 we discussed how the hormone oxytocin, released when people are in love, may be instrumental in producing a long-term change in behaviour. In some circumstances, that change may have facilitated the prolonged commitment by both partners to raising children. Indeed, the biological benefit of such commitment may well have been the reason why such a mechanism, with its subjective states of intense emotion, evolved in the first place.

The variety of human marriage institutions probably reflects a multitude of economic and social pressures.[37] It is a mistake, we believe, to push too strongly arguments that are derived purely from considerations of biological design. But what about the famous case, often taken as a justification for linking biology with the emergence of human custom – the widespread taboo against incest?

From Inbreeding Avoidance to Incest Taboos

At the back of the Anglican *Book of Common Prayer* is a Table of Kindred and Affinity, 'wherein whosoever are related are forbidden by the Church of England to marry together'. The table stipulates that a man may not marry his mother, sister, daughter, or a variety of other related individuals. Neither may a woman marry an uncle, nephew, grandparent or grandchild. These elaborate and highly specific rules are typical of incest taboos found in virtually all human cultures, and at first glance they look as though they are based on good biological sense because they seem to have been formulated to minimise the costs of inbreeding. But the list continues with some more surprising exclusions. For example, a man may not marry his wife's father's mother or his daughter's son's wife. The mind boggles. What was life like in the sixteenth century if the Church felt it necessary to include such prohibitions? And did women really live long enough to even contemplate marrying their deceased granddaughter's husband?

At least six of the twenty-five relationships in the table that are precluded from marriage involve no genetic link at all, and several are ambiguous on this point. Interestingly, though, the Church of England did not worry about marriages between first cousins. Other cultures do, but here again striking inconsistencies are found. In many cultures marriages between parallel first cousins are forbidden, whereas marriages between cross-cousins are allowed and even actively encouraged. (A parallel-cousin is the child of the father's brother or the mother's sister; a cross-cousin is the child of the father's sister or the mother's brother.) To a biologist interested in the functions of such incest taboos, this distinction is puzzling, because parallel- and cross-cousins, despite being treated so differently in the marriage rules of many cultures, do not differ genetically.

The term 'incest taboo', which refers to a culturally transmitted prohibition in humans, is not the same thing as the mating inhibition seen in other species. Even so, the possible links between biological evolution and cultural history are intriguing. Humans, like many other animals, are generally disinclined to have a prolonged sexual relationship with individuals of the opposite sex whom they have known from an early age – even when such a relationship is encouraged. The developmental process giving rise to this reluctance is presumably a product of biological evolution. But how (if at all) was it linked to a similar prohibition transmitted by word of mouth and then, much later, written down in social rules?

In a book published at the end of the nineteenth century, Edward Westermarck suggested that humans have an inclination to prevent other people behaving in ways in which they would not themselves behave.[38] On this view, left-handers were in the past forced to adopt the habits of right-handers because the right-handers found them disturbing. In the same way, those who were known to have had sexual intercourse with close kin were discriminated against. People who had grown up with kin of the opposite sex were generally not attracted to those individuals, and disapproved when they discovered others who were. It was nothing to do with society not wanting to look after the half-witted children of inbreeding, since in many cases they had no idea that inbreeding was the cause. Rather, the disapproval was about suppressing abnormal behaviour which is potentially disruptive in small societies. Such conformity looks harsh to modern eyes, even though we have plenty of examples of it in contemporary life. However, when so much depended on unity of action in the environment in which humans evolved, wayward behaviour could have destructive consequences for everybody. It is not difficult to see why conformity should have become a powerful trait in human social behaviour.

Once in place, the desire for conformity on the one hand, and the reluctance to inbreed on the other, would have combined to generate social disapproval of inbreeding. The emergence of incest taboos would take on different forms, depending on which sorts of people, non-kin as well as kin, were likely to be familiar from early life. Familiarity accounts for the strange proscriptions in the Church of England's marriage rules. The explanation works well for the societies in which parallel-cousins are prohibited from marrying but marriages between cross-cousins are favoured. In these cultures, the parallel-cousins tend to grow up together because brothers tend to stay with brothers and sisters with sisters, whereas the cross-cousins are generally less familiar with each other because brothers and sisters tend to live in different places after marriage.

If these ideas are correct, human incest taboos did not arise historically from deliberate intention to avoid the biological costs of inbreeding. Rather, in the course of biological evolution, two separate mechanisms appeared. One was a developmental process concerned with striking an optimum balance between inbreeding and outbreeding when choosing a mate. The other was concerned with social conformity. When these two propensities were put together, the result was social disapproval of those who chose partners from within their close family. When social disapproval was combined with language, verbal rules appeared which could be transmitted from generation to generation, first by word of mouth and later in written form. If this view of the historical origins of incest taboos is correct, the traditional sharp distinction between biology and culture is less than helpful. Once again, understanding comes from knowledge of how behaviour develops in individuals.

13

This Strange Eventful History

Man's life is well comparèd to a feast,
Furnished with choice of all variety.
Richard Barnfield, 'Man's Life' (1598)

The One True Cause

The effectiveness of education, the role of parents in shaping the characters of their children, the causes of violence and crime, and the roots of personal unhappiness are self-evidently matters of huge importance. And, like so many other fundamental issues about human existence, they all relate to behavioural development. The catalogue continues. Do bad experiences in early life have a lasting effect? Is intelligence in the genes? Can adults change their attitudes and behaviour? When faced with such questions, most people want simple answers. They want to know what *really* makes the difference.

We have argued that looking for single causes, whether genetic or environmental, may yield answers of a kind, but little sense of what happens as each individual grows up. The language of nature versus nurture, or genes versus environment, gives only a feeble insight into the processes which we have called developmental cooking. The best that can be said of the nature/

nurture split is that it provides a framework for uncovering a few of the genetic and environmental ingredients which generate differences between people. At worst, it satisfies a demand for simplicity in ways that are fundamentally misleading.

The search for simple environmental origins, which had wide appeal in the mid-twentieth century, has been partly superseded by an equally skewed belief in the overriding importance of genes. If pressed, scientists may concede that their talk of genes 'for' shyness, maternal behaviour, promiscuity, verbal ability, criminality, or whatever, is merely a shorthand. They may, however, try to legitimise the language of genes 'for' behaviour, by pointing to seemingly straightforward examples like the genes for eye colour. Nonetheless, the notion of genes 'for' behaviour undoubtedly corrupts understanding.

A single developmental ingredient, such as a gene or a particular form of experience, might produce an effect on behaviour, but this certainly does not mean that it is the only thing that matters. Even in the case of eye colour, the notion that the relevant gene is *the* cause is misconceived, because all the other genetic and environmental ingredients that are just as necessary for the development of eye colour remain the same for all individuals. A more honest translation of the 'gene for' terminology would be something like: 'We have found a particular behavioural difference between individuals which is associated with a particular genetic difference, all other things being equal.' The media and the public might start to get the message if plain language like this were used routinely.

The common image of a genetic blueprint for behaviour also fails because it is too static, too suggestive that adult organisms are merely expanded versions of the fertilised egg. In reality, developing organisms are dynamic systems that play an active role in their own development. Even when a particular gene or a particular experience is known to have a powerful effect on the development of behaviour, biology has an uncanny way of finding alternative routes. If the normal developmental pathway to a particular form of adult behaviour is impassable, another way may often be found. The individual may be able, through its behaviour, to match its environment to suit its own characteristics – a process dubbed 'niche-picking'. At the same time, playful activity increases the range of available choices and, at its most creative, enables the individual to control the environment in ways that would otherwise not be possible.

The low-tech cooking metaphor, which we favour, serves to shift the focus onto the multi-causal and conditional nature of development. Using butter instead of margarine may make a cake taste different when all the other ingredients and cooking methods remain unchanged. But if other combinations of ingredients or other cooking methods are used, the distinctive difference between a cake made with butter and a cake made with margarine may vanish. Similarly, a baked cake cannot readily be disaggregated into its original raw ingredients and the various cooking processes, any more than a behaviour pattern or a psychological characteristic can be disaggregated into its genetic and environmental influences and the developmental processes that gave rise to it. But to pursue this metaphor, certain ingredients of a cake, such as the raisins, may recognisably survive most styles of baking. In an analogous way, certain aspects of human behaviour, such as the facial expression of emotions, are found in every culture.

To use a different metaphor, development is not like a fixed musical score, which specifies exactly how the performance starts, proceeds and ends. It is more like a form of jazz in which musicians improvise and elaborate their musical ideas, building on what the others have just done. As new themes emerge, the performance acquires a life of its own, and may end up in a place none could have anticipated at the outset. Yet it emerges from within a fixed set of rules and the constraints imposed by the musical instruments.

It no longer seems obscure to talk about systems and processes. This change in thinking may reflect commonplace experience. From an early age many people are exposed to computer games, in which outcomes depend on a combination of conditions. Children playing such games meet, for instance, in the dungeon of the dark castle a dragon that can only be killed with a special sword which had to picked up on the top of the crystal mountain. In doing so, they have begun to accustom themselves to the contextual, conditional character of the real world. The linear thinking of a previous generation, with every event having a single cause, is slowly being replaced by an understanding of co-ordinated process.

The Evolution of Development

The processes of behavioural and psychological development that give rise to an individual's behaviour are themselves the products of Darwinian evolution, and look as though they have been designed. The design features of development are evident, for example, in sensitive periods. Certain kinds of learning, such as learning about language, occur at a particular stage in development, usually early in life. The developmental processes that make learning easier at the beginning of a sensitive period are often linked to physical growth. But they are also timed to correspond with changes in the ecology of the developing individual. The processes that bring the sensitive period to an end are often related to the gathering of crucial information, such as the physical appearance of the individual's mother or close kin. Consequently, these processes normally do not terminate the sensitive period until that information has been gathered. In the unpredictable real world, the age when the individual can acquire crucial knowledge is variable; the design of the developmental process reflects that uncertainty.

The way in which well-designed behaviour evolved is the province of Darwinian theory, which provides the most coherent evolutionary explanation for biological design. But does the systems character of development threaten Darwinian theory? Is the link between gene and behaviour too indirect, and the causal pathway too long, to allow an evolutionary process acting on the end products of development to change gene frequencies? We think not. Animal breeders are manifestly able to change specific aspects of behaviour in a heritable way, by means of artificial selection. Genetic differences clearly do give rise to specific behavioural differences.

The evolutionary process does not require a simple correspondence between genes and adaptive behaviour. Darwinian evolution operates on individuals that have developed within a particular set of conditions. If those conditions are stable for many generations then the changes that matter will be primarily genetic. Individuals vary; some survive and reproduce more successfully than others because they possess a crucial characteristic; and close relatives are more likely to share that characteristic than unrelated individuals. Apparent design is produced, even when it is at the end of the long and complicated process of development. But the environment does not cease to be important for evolution just because it remains constant. Change the environment and the outcome of an individual's development may be utterly different. Indeed, if an individual does not inherit its parents' environment along with their genes, it may not be well adapted to the conditions in which it now finds itself.

It is, alas, all too obvious that humans do stupid things which run counter to their best interests. John Bowlby, who pioneered a biological approach to behavioural development, was well aware that some adaptations which benefited humans in the past have become dysfunctional in the radically different modern world.[1] Behaviour such as seeking out and receiving pleasure from eating sweet or fatty foods was doubtless vital in a subsistence environment, but in a well-fed society it does more harm than good.

In other respects, though, humans have shown remarkable capacity for rapid change. The transformation of man-made environments, and subsequent human adaptations to them, have been abrupt and recent, relative to human evolutionary history. The earliest forms of civilisation, in the shape of systematic farming, emerged less than 10,000 years ago. The first written records appeared 6,000 years ago in Mesopotamia and China, and the wheel was invented not long afterwards. Industrialised societies started to emerge within the past 200 years and computers became ubiquitous in the later part of the twentieth century. The changes in environmental conditions from those in which humans evolved have been radical and have occurred when genetic change has been negligible. Human genes have changed little since humans emerged in their present form about 100–150,000 years ago. Given the rate at which genes are known to alter through mutation, less than 0·1 per cent of human genetic material has changed over this period. A hunter-gatherer baby born on the plains of Africa thousands of years before the wheel was invented, and transported miraculously through time to the present, would have been fundamentally no different in terms of genetic make-up from his or her modern counterpart. Given the right circumstances, the hunter-gatherer baby could in principle have grown up to become a lawyer, a taxi driver or a molecular biologist.

Opportunism also plays an important role in driving evolutionary change. As in a large restaurant, many different dishes are being cooked at the same time in the kitchen of behavioural development. Occasionally the behavioural dishes are thrown together and something quite novel – and useful – is produced by chance. Humans are perfectly capable of appreciating the value of their own experiments, and the emerging effects have had an extraordinary influence on human history. The combination of spoken language, which has obvious utility in its own right, and manual dexterity in fashioning tools, which also has obvious utility, combined at a particular and relatively recent moment in evolutionary history to generate written language. The discovery of written language took place several times and in several forms in different parts of the world, with ideas represented by pictures or spoken sounds represented by symbols. Once invented, the techniques were quickly copied and became crucial elements of modern civilisation. It was that active combining of different capacities which started the whole remarkable cultural sequence of events.

Who Chooses, Who Designs?

'Design for a life' is not intended to refer to conscious planning. The developmental processes are seemingly designed by Darwinian evolution, but the development of an individual organism need not involve much conscious thought or planning. In theory, apparent choice need reflect nothing more than a bias predetermined by evolutionary or personal history. People may choose a course of action without knowing why or reflecting a great deal on what they do. The behaviour of brain-damaged patients reveals how crucial the emotions are in real-life decision-making. Non-conscious biases advantageously guide behaviour before conscious knowledge or rational analysis come into operation. But this leaves a nagging question. When people make what seem like conscious choices, do they really know what they are doing?

Evolutionary psychology might appear to imply that individuals do not make free choices. A well-designed life cycle, with its specialisations needed for each stage, its responsiveness to local conditions, its developmental scaffolding, and its rules for assembly, is the result of evolutionary processes. Intentions do not come into it. And yet, individuals clearly do make a big difference to what happens in their lives through their choices and decisions. They may be surprised by the consequences of their own actions. A well-designed brain should be able to anticipate the consequences of various courses of action and choose between them on the basis of their likely costs and benefits. Planning before doing is clearly of great advantage. It is obvious that people often do make sensible, well-considered decisions. Their decisions may reflect developmental mechanisms that are themselves the results of evolutionary processes, but it would be foolish to look for adaptation in every individual choice. A proper understanding of biology brings back free will – even if the freedom is constrained and sometimes used unwisely.

What sets the biases, providing the basis for a well-adjusted emotional life and thence rational decision-making? Much remains unknown, but evidence grows for the pre-emptive effects of certain experience, some of it occurring before birth. Individuals have the potential to play a number of different developmental tunes, but in the course of their lives play only one of them. The particular tune that is played may be triggered by a feature of the environment in which that individual is growing up, and is adapted to the conditions in which it is played. A short-lived organism could benefit from being equipped with a jukebox-like developmental mechanism to set it on the right track for the particular environment it is going to inhabit. If the animal only lives for a few months, then cues received from its mother while she is pregnant may well provide a reliable prediction of the world into which it is going to be born. But how could such a developmental mechanism ever have evolved in slowly developing, long-lived animals like humans, for whom long-range forecasts are likely to be of limited value? The world inhabited by an adult might be totally different from the one in which the individual's mother lived. Why not just adapt to new conditions as they arise during the life cycle?

One answer may be that physiological processes for dealing with particular climates and diets are not easily reorganised. If so, forecasting would be beneficial for animals who were born into particular conditions like the Arctic, deserts, or thick jungle, and who were likely to stay in those environments all their lives – as would commonly have been the case until recently in human history. The environmental forecast would have been much more reliable in the past than it could be now, in a rapidly changing industrialised society. And it would be most beneficial if the triggering occurred in early life, within a sensitive period, when the optimal metabolism, body structure and behavioural repertoire could be established most readily.

When attempting to understand development it is crucial to remember that, in the course of biological evolution, the possibilities for throwing away existing designs and starting again from scratch became progressively more difficult as organisms became ever more complicated. Successful modifications are usually grafted onto the end of a long developmental process. As a consequence, biological systems have features that a tidy-minded human engineer would hate. The vertebrate eye is a good example. The light-sensitive cells lie behind the connections through the optic nerve to the brain, so the image of the external world has to be focused onto the retina through these connections. Parallels with human engineering do exist, however. Makers of elaborate objects like cars do not casually discard their most expensive machine tools. Consequently, their designs often conserve features of older cars that would have been discarded if the whole machine had been designed afresh each time.

Building Clever Machines

The slow process of development that is seen in animals presumably evolved because the complexity of adult behaviour, especially in humans, cannot be assembled adequately without a childhood. If the design principles of behavioural development are truly powerful, they should have broader application in the non-biological world. If a long period of development is necessary to build a complex, intelligent organism, then the same may well be true of intelligent machines – even if their physical structures have little in common with living beings.

Most designers of intelligent machines have built the equivalents of adults – machines that can learn, but are otherwise fully formed at 'birth'. However, the implication from biology is that a different approach may have to be taken, one in which the machine is 'born' as a baby and undergoes a long period of development. In more recent times, various schemes for creating intelligent robots have started to exploit some of the design principles of development, by giving the machine a form of childhood in which it develops through interacting with humans and its physical environment.

The design features of biological development have other interesting implications when considered in the context of intelligent machines. Animals and people can modify their behaviour with increasing elaboration on the basis of their experience, but they also tend to settle into familiar habits that restrict their capacity for further change. Sometimes it is necessary to break free from these restrictions. General-purpose, intelligent machines may also have to deploy tactics to deal with new situations that could not have been foreseen when they were designed. One important element of biological development is play. Through play, young animals and humans acquire valuable experience, knowledge and skills that enable them to adapt to the particular circumstances in which they will have to live. The same will probably be true of highly intelligent machines. When asked to solve problems, a machine may find an answer that is not the best one. It will be better able to arrive at an optimal solution if it is designed to engage in the computer equivalent of play. Having found one solution, the machine casts around in case a better solution can be found. It tries out novel combinations of actions, even though many will land up in blind alleys. The cost is worth it if the machine discovers a novel and efficient solution which it can deploy for the rest of its working life.

The child is not a miniature adult, nor is it necessarily incompetent. Many of the behavioural and physiological features of young organisms are required for survival when they are much smaller than adults, and are forced by their size and dependence to live in a quite different habitat. Babies and children have special faculties and behaviour patterns that are suited to their habitat at that period in development, and which are not necessarily precursors of adult behaviour. Children have abilities that adults lack and, as a consequence, see things in different ways. As the developmental psychologist William Kessen has pointed out, development is not all about progress.[2] By analogy, special faculties may be needed when an elaborate structure such as an intelligent machine is developing. These faculties do not have to remain in place for ever and may be removed, just as scaffolding is eventually removed from a building. General purpose, intelligent machines may need to be dependent on a care-giver while they are acquiring crucial information during a sensitive period early in their working lives. And machines equipped with the capacity to adopt one of several distinct modes of behaviour, triggered by prevailing conditions early on, may be said to have the equivalent of the alternative lives available to developing animals.

A Process Too Complicated to Explain?

For all their individuality and ingenuity, humans cannot escape the inevitable progression of what Shakespeare called 'this strange eventful history', through the seven ages of life – or eight ages when the period before birth is included. Every person is stuck with the basic life cycle of the human species and must cope with many different problems during his or her development. Like buildings, bodies are difficult to alter fundamentally once they are fully constructed. Such constraints generate continuities in behaviour as well as physical structures. Moreover, people tend to settle into familiar habits and thereby restrict further opportunity for change. The conditions in which change happens may sometimes involve stress, but the capacity to change in adulthood is nonetheless real. For the old, the rediscovery that they can still change bears out the point that development continues until death.

Modern biology does not support common prejudices about the inability of humans to change. Aggression and violence, for example, are depressingly widespread and frequently gratuitous. But they are also mutable, as are the human institutions of war that harness them.[3] Despite the shameful catalogue of wars and violence that have darkened human history, civilisation and society are demonstrably capable of creating conditions in which the great majority of people behave peaceably towards their fellow humans for long periods. It is crucial to have a good understanding of those conditions. The same point about understanding applies to the treatment of childhood.

In their ambition for their children, or their anxiety about real and imagined dangers, some parents underestimate the importance of the seemingly pointless aspects of childhood, the time when their children can roam freely with friends and discover things in their own way and at their own pace. Children who are pushed too hard academically, and who consequently seem to advance beyond their peers, may ultimately pay a price in terms of lost opportunities for development. Equally, their childhood may be foreshortened if their parents pay too much attention to promoting musical, athletic or other specialist skills at the expense of everything else. Childhood is, amongst other things, a period of intense socialisation, and children benefit from a rich and varied diet of experience, not just success in exams and competitions – important though these undoubtedly are. Through an excessive emphasis by some parents on certain forms of achievement, their children end up with narrow lives. The long-term effects on their social, emotional and intellectual development may ironically result in poorer attainment later in life, as well as personal unhappiness. A faster, narrower development is not a sure-fire recipe for success.

This is not a new thought. More than two hundred years ago Goethe wrote a short musical play called *Erwin and Elmira*. At one point Olympia tells her daughter Elmira that the freedom and joys of childhood have been lost to the 'young' generation, preoccupied as they are with clothes and etiquette:

> We played and jumped and racketed about, and as quite big girls we still enjoyed swings and ball games . . . We ran around the house together in our plainest clothes, playing silly games for silly prizes, gloriously happy.

Childhood has important functions. Nevertheless, a criticism of adults who push children too hard is not a plea for a release from all discipline. Children need boundaries and direction. They must grow up to function in a demanding world in which they will require complex skills, some of which can be acquired only with perseverance and difficulty. Parental care and professional teaching skills are more important than ever, but they have to be properly tuned to the individual child's stage of development and personal characteristics. Anyone with any experience of children knows that they are all unique. Children learn in different ways, which skilful teachers recognise as they try, if they have time, to coax the best out of each individual.

The simple message from biology for children and their parents is: do not be in too much of a hurry. Childhood has a function. In the words of a nineteenth-century novelist: 'Many a genius has been slow of growth. Oaks that flourish for a thousand years do not spring up into beauty like a reed.' Children should be given the time, the stimulation and the support they need to develop all their varied attributes, bringing these into balance and understanding the competition and co-operation of the social world that surrounds them. Such an attitude properly respects the life for which childhood was designed by evolutionary processes. But it also recognises the reality of the present-day world in which children must function with sensitivity and skill. The best gift that can be given to a child is the happiness that comes from being able to cope successfully in a complex world.

We have aimed to give some understanding of what happens as an individual grows up: how each human comes to be the way he or she is, and why. We have explored the ways in which genes and experience come together in the developmental kitchen and contribute to the process of development; how each individual has the capacity to develop in a number of distinctly different ways; how each individual's characteristics start to emerge early in life and yet remain capable of transformation; how early experiences affect sexual preferences later in life; and how both chance and choice affect each individual's life.

Old styles of thought die hard. Even the much-used distinction between sex and gender has a strong whiff of the old genes/environment opposition. 'Sex' is biology; but 'gender' is part of culture, the acquired behaviour deemed appropriate to the social role of that sex. Any fair-minded person listening to set-piece debates about behavioural development is likely to be left irritated by the claims and counter-claims. And well-meaning attempts to break out of the nature–nurture straitjacket have often resulted in an obscure and bewildering portrayal of development as a process of impenetrable complexity (what Salman Rushdie once described in another context as a P2C2E – a Process Too Complicated To Explain). Indeed, development seemed so unfathomably complex to eighteenth-century biologists that they believed that it must depend on supernatural guidance.

The processes involved in behavioural development do indeed look forbiddingly complicated on the surface. Some would argue that it is worse underneath and that such order as is found is generated by dynamic systems of great complexity.[4] In contrast, we have suggested that simplicity and regularity can be found in the developmental processes that give rise to unique individuals. Confidence in that conclusion comes not from general principles but from the careful analysis of particular cases. The essence of development – change coupled with continuity – starts to make sense. It becomes possible to understand how the individual is so responsive to events at one stage and so unaffected by them at another. After much rummaging in the kitchen cupboard, out are coming some of the critical ingredients for a well-designed life.

Acknowledgements

Some very kind people read parts or all of the book in draft and we are extremely grateful to them. The mistakes are all ours, however. We warmly thank: Max Alexander, Dusha Bateson, Gillian Bentley, Bryn Caless, Nic Coombs, Philip Davies, Ian Donaldson, Nick Hales, Robert Hinde, Barry Keverne, Phyllis Lee, Brendan Lehane, Charlie Loke, Aubrey Manning, Harriet Martin, Susan Oyama, Bob Rowthorn, John Sants, Barbara Herrnstein Smith, Patsy Wilkinson, Bernard Williams and Michael Yudkin.

The authors and publishers are grateful to the following for permission to reproduce extracts from works in copyright: extract from 'This Be the Verse' by Philip Larkin by permission of Faber and Faber; extracts from *All Said and Done* by Simone de Beauvoir, translated by Patrick O'Brian, by permission of the Estate of Simone de Beauvoir and Penguin Books Ltd; extract from *The New Confessions* by William Boyd by permission of Penguin Books Ltd; extract from *Anna Karenina* by Leo Tolstoy, translated by Rosemary Edmonds, by permission of Penguin

Books Ltd; extract from *Conundrum* by Jan Morris by permission of Peters Fraser & Dunlop Group Ltd; extract from *Now We Are Six* by A.A. Milne by permission of Egmont Children's Books; extract from *Cold Comfort Farm* by Stella Gibbons by permission of Curtis Brown Group Ltd; extract from *Nineteen Eighty-Four* by George Orwell by permission of A.M. Heath & Co Ltd on behalf of the Estate of the Late Sonia Brownell Orwell, and Martin Secker & Warburg Ltd; extract from *Fifth Philosopher's Song* by Aldous Huxley, permission sought through Random House; extract from *On the Black Hill* by Bruce Chatwin, permission sought through Random House; extract from 'Big Two-Hearted River' by Ernest Hemingway by permission of The Hemingway Foreign Rights Trust; extract from *Captain Corelli's Mandolin* by Louis de Bernières, originally published by Secker & Warburg, by permission of Random House; extract from *The Prophet* by Kahlil Gibran, permission sought through Random House; extract from *Room at the Top* by John Braine, originally published by Eyre & Spottiswoode, by permission of David Higham Associates; extract from *The Dyers Hand* by W.H. Auden by permission of Faber and Faber; extract from *Peter Pan and Wendy* by J.M. Barrie by permission of the Great Ormond Street Hospital Charity, 1937; extract from 'Origins of Society' by Ernest Gellner in *Origins*, edited by A.C. Fabian, by permission of Cambridge University Press; extract from *Translation from the Natural World* by Les Murray by permission of Carcanet Press Ltd; extracts from *Battle for the Mind* by W. Sargant, permission sought from A.P. Watt.

Every reasonable effort has been made to contact copyright holders for all the extracts reproduced in this volume. However, it has not been possible to make contact with all copyright holders. The authors and publishers would ask, therefore, that any copyright holder who feels that a quotation contained herein may contravene their copyright contact Jonathan Cape at 20 Vauxhall Bridge Road, London SW1V 2SA.

References

We have been sparing in our use of references since this is not an academic book. But we have included a reference when we anticipated that a curious or sceptical reader might ask, 'How do they know that?' We have not always cited the original source, but have instead tried to choose recent reviews or books that provide a good point of entry into the relevant scientific literature on a subject.

1 The Developmental Kitchen

1 Barlow N. (ed.), *The Autobiography of Charles Darwin* (Collins, 1958).
2 Figes, O., *A People's Tragedy* (Jonathan Cape, 1996).
3 Gould, S.J., and Vrba, E., *Palaeobiol.*, vol. 8 (1982), p. 4.

2 The Seven Ages

1 Burrow, J.A., *The Ages of Man* (Oxford University Press, 1986).
2 Stearns, S.C., *The Evolution of Life Histories* (Oxford University Press, 1992).
3 Trivers, R.L., *Am. Zool.*, vol. 14 (1974), p. 249.
4 Haig, D., *Q. Rev. Biol.*, vol. 68 (1993), p. 495.

5 Renfree, M.B., in Austin, C.R., and Short, R.V. (eds.), *Reproduction in Mammals*, vol. 2, *Embryonic and Fetal Development* (Cambridge University Press, 1982), p. 26.
6 Gottlieb, G., in Aronson, L.R. et al. (eds.), *Development and Evolution of Behavior* (W.H. Freeman, 1970), p. 111.
7 Fifer, W.P., and Moon, C.M., *Acta Paediat. Suppl.*, vol. 397 (1994), p. 86.
8 DeCasper, A.J. and Fifer, W.P., *Science*, vol. 208 (1980), p. 1174.
9 Pendergast, M., *Victims of Memory* (HarperCollins, 1996).
10 Hall, W.G., and Williams, C.L., *Adv. Study Behav.*, vol. 13 (1983), p. 219.
11 Dechateau, P., *Acta. Paediat. Scand.*, S344 (1988), p. 21.
12 Rutter, M., *J. Child Psychol. Psychiat.*, vol. 36 (1995), p. 549.
13 Murray, L., and Trevarthen, C., *J. Child Lang.*, vol. 13 (1986), p. 15.
14 Meltzoff, A.N., and Moore, M.K., *Science*, vol. 198 (1977), p. 75.
15 Darwin, C., *Mind*, vol. 2 (1877), p. 285.
16 Kessen, W., and Scott, D.T., in Levine, M.D., et al. (eds.), *Developmental-Behavioral Pediatrics* (W.B. Saunders, 1992), p. 3.
17 Boden, M.A., *Piaget* (Harvester Press, 1979).
18 Bateson, P., *Trends Ecol. Evol.*, vol. 9 (1994), p. 399.
19 Pettit, G.S., and Bates, J.E., *Dev. Psych.*, vol. 25 (1989), p. 413.
20 Mock, D.W., and Parker, G.A., *The Evolution of Sibling Rivalry* (Oxford University Press, 1997).
21 Mock, D.W., and Parker, G.A., *Anim. Behav.*, vol. 56 (1998), p. 1.
22 Bohler, E., and Bergstrom, S., *J. Trop. Pediat.*, vol. 42 (1996), p. 104.
23 Dunn, J., and Plomin, R., *Separate Lives* (Basic Books, 1990).
24 Harris, J., *The Nurture Assumption* (Bloomsbury, 1998).
25 Offer, D., and Schonert-Reichl, K.A., *J. Am. Acad. Child Adolesc. Psychiat.*, vol. 31 (1992), p. 103.
26 Daly, M., and Wilson, M., *Homicide* (de Gruyter, 1988).
27 Office for National Statistics, *Social Trends* 27 (Stationery Office, 1997).
28 Lamberts, S.W.J., et al., *Science*, vol. 278 (1997), p. 419.
29 Clutton-Brock, T.H., *The Evolution of Parental Care* (Princeton University Press, 1991).

30 Hawkes, K.K., et al., *Proc. Natl. Acad. Sci. USA*, vol. 95 (1998), p. 1336.
31 Oudshoorn, N.E., *J. Psychosom. Obstet. Gynaecol.*, vol. 18 (1997), p. 137.
32 Schow, D.A., et al., *Postgrad. Med.*, vol. 101 (1997), p. 62.
33 Bulpitt, C.J., et al., *Aging (Milano)*, vol. 6 (1994), p. 181.
34 Williams, G.C., *Plan and Purpose in Nature* (Weidenfeld & Nicolson, 1996).
35 Ricklefs, R.E., *Am. Nat.*, vol. 152 (1998) p. 24.
36 Goto, M., *Mech. Ageing Dev.*, vol. 98 (1997) p. 239.
37 Gray, M.D., et al., *Nature Genet.*, vol. 17 (1997), p. 100.

3 *The Sparks of Nature*

1 Constancia, M., et al., *Genome Res.*, vol. 8 (1998), p. 881.
2 Veytsman, B.A., *Evol. Ecol.*, vol. 11 (1997), p. 519.
3 Sokolowski, M.B., et al., *Proc. Natl. Acad. Sci. USA*, vol. 94 (1997) p. 7373.
4 Skuse, D.H., et al., *Nature*, vol. 387 (1997), p. 705.
5 Lummaa, V., et al., *Nature*, vol. 394 (1998), p. 533.
6 Jablonka, E., and Lamb, M.J., *Epigenetic Inheritance and Evolution* (Oxford University Press, 1995).
7 Devlin, B., et al., *Nature*, vol. 388 (1997), p. 468.
8 Wright, L., *Twins* (Weidenfeld & Nicolson, 1997).
9 Bouchard, T.J., et al., *Science*, vol. 250 (1990), p. 223.
10 Wilson, R.S., *Child Dev.*, vol. 54 (1983), p. 298.
11 Mackintosh, N.J., *IQ and Human Intelligence* (Oxford University Press, 1998).
12 Schweinhart, L.J., et al., *Significant Benefits* (High/Scope Press, 1993).
13 Yoshikawa, H., *Fut. Child.*, vol. 5 (1995), p. 51.
14 Schweinhart, L.J., and Weikart, D.P., *Lasting Differences* (High/Scope Press, 1997).
15 Levine, S., et al., *Ann. NY Acad. Sci.*, vol. 807 (1997), p. 210.
16 van Oers, H.J.J., et al., *Endocrinol.*, vol. 139 (1998) p. 2838.
17 Anisman, H., et al., *Int. J. Dev. Neurosci.*, vol. 16 (1998), p. 149.
18 Liu, D., et al, *Science*, vol. 277 (1997), p. 1659.
19 Sulloway, F.J., *Born to Rebel* (Little, Brown, 1996).

20 Strachan, D.P., et al., *J. Allerg. Clin. Immunol.*, vol. 99 (1997), p. 6.
21 Zajonc, R.B., *Psychol. Bull.*, vol. 93 (1983), p. 457.
22 Schubert, D.S., et al., *J. Psychol.*, vol. 95 (1977), p. 147.
23 Sulloway, F.J., *Born to Rebel* (Little, Brown, 1996).
24 Blanchard, R., et al., *Arch. Sex. Behav.*, vol. 25 (1996), p. 495.
25 Zadnik, K., *Optom. Vis. Sci.*, vol. 74 (1997), p. 603.
26 O'Connor, N., *Br. J. Disord. Commun.*, vol. 24 (1989), p. 1.
27 Barwick, J., et al., *Br. J. Educ. Psychol.*, vol. 59 (1989), p. 253.
28 Sloboda, J., *Nature*, vol. 362 (1993), p. 115.
29 Coon, H., and Carey, G., *Behav. Genet.*, vol. 19 (1989), p. 183.
30 Baharloo, S., et al., *Am. J. Hum. Genet.*, vol. 62 (1998), p. 224.
31 Krampe, R.T., and Ericsson, K.A., *J. Exp. Psychol. Gen.*, vol. 125 (1996), p.331.
32 Plomin, R., *Science*, vol. 248 (1990), p. 183.
33 Plomin, R., et al., *Behavioral Genetics* (W.H. Freeman, 1997).
34 Cooper, R.M., and Zubek, J.P., *Can. J. Psychol.*, vol. 12 (1958), p. 159.

4 *Cooking Behaviour*

1 Coen, E., *The Art of Genes* (Oxford University Press, 1999).
2 Wu, C.S., et al., *Proc. Natl. Acad. Sci. USA*, vol. 75 (1978), p. 4047.
3 Kauffman, S.A., *The Origins of Order* (Oxford University Press, 1993).
4 Pfaff, D.W., *Proc. Natl. Acad. Sci. USA*, vol. 94 (1997), p. 14213.
5 Kear, J., *Proc. Zool. Soc. Lond.*, vol. 138 (1962), p. 163.
6 Sulloway, F.J., *Born to Rebel* (Little, Brown, 1996).
7 Scarr, S., and McCartney, K., *Child Dev.*, vol. 54 (1983), p. 424.
8 Stabler, B., et al., *J. Dev. Behav. Pediat.*, vol. 15 (1994), p. 1.
9 Hinde, R.A., *Relationships* (Psychology Press, 1997).
10 Fox, P.W., et al., *Nature*, vol. 384 (1996), p. 356.
11 Cloninger, C., et al., *Arch. Gen. Psychiat.*, vol. 39 (1982), p. 1242.
12 Mackintosh, N.J., *IQ and Human Intelligence* (Oxford University Press, 1998).
13 Bateson, P., *Int. J. Behav. Dev.*, vol. 10 (1987), p. 1.
14 Shields, J., *Monozygotic Twins Brought Up Apart and Brought Up Together* (Oxford University Press, 1962).
15 Dunn, J., and Plomin R., *Separate Lives* (Basic Books, 1990).

16 Rutter, M., and Rutter, M., *Developing Minds* (Basic Books, 1993).
17 Zajonc, R.B., *Psychol. Bull.*, vol. 93 (1983), p. 457.
18 Tseng, W.S., et al., *Am. J. Psychiat.*, vol. 145 (1988), p. 1396; Falbo, T., and Poston, D.L., *Child Dev.*, vol. 64 (1993), p. 18; Wan, C.W., et al., *J. Genet. Psychol.*, vol. 155 (1994), p. 377.
19 Standing, L., *Q. J. Exp. Psychol.*, vol. 25 (1973), p. 207.

5 *Protean Instincts*

1 Bateson, P., in Magnusson, D. (ed.), *The Lifespan Development of Individuals* (Cambridge University Press, 1995), p. 1.
2 Darwin, C., *Mind*, vol. 2 (1877), p. 285.
3 Vollrath, F., et al., *Physiol. Behav.*, vol. 62 (1997), p. 735; Krink, T., and Vollrath, F., *Proc. Roy. Soc. Lond. B*, vol. 265 (1998), p. 2051.
4 Pinker, S., *The Language Instinct* (Penguin, 1994).
5 Nisbett, A., *Konrad Lorenz* (Dent, 1976).
6 Lorenz, K., *Evolution and Modification of Behavior* (University of Chicago Press, 1965).
7 Grohmann, J., *Z. Tierpsychol.*, vol. 2 (1939), p. 132.
8 Gwinner, E., *J. Exp. Biol.*, vol. 199 (1996), p. 39.
9 Freedman, D.G., *J. Child Psychol. Psychiat.*, vol. 5 (1964), p. 171.
10 Bruce, V., and Young, A., *In the Eye of the Beholder* (Oxford University Press, 1998).
11 Haxby, J.V., et al., *Proc. Natl. Acad. Sci. USA*, vol. 93 (1996), p. 922.
12 Ekman, P., et al., *J. Pers. Soc. Psychol.*, vol. 53 (1987), p. 712.
13 Lehrman, D.S., *Q. Rev. Biol.*, vol. 28 (1953), p. 337.
14 Beach, F.A., *Psychol. Rev.*, vol. 62 (1955), p. 401.
15 Lehrman, D.S., in Aronson, L.R., et al. (eds.) *Development and Evolution of Behavior* (W.H. Freeman, 1970), p. 17.
16 Bateson, P., in Bateson, P. (ed.), *The Development and Integration of Behaviour* (Cambridge University Press, 1991), p. 19.
17 Wilson, D.S., *Ethol. Sociobiol.*, vol. 15 (1994), p. 219.
18 Berridge, K.C., in Hogan, J.A., and Bolhuis, J.J. (eds.), *Causal Mechanisms of Behavioural Development* (Cambridge University Press, 1994), p. 147.
19 Cowey, A., and Weiskrantz, L., *Neuropsychologia*, vol. 13 (1975), p. 117.

20 Gottlieb, G., *Development of Species Identification in Birds* (University of Chicago Press, 1971).
21 Martin, P., and Caro, T.M., *Adv. Study Behav.*, vol. 15 (1985), p. 59.
22 Hailman, J.P., *Behaviour Suppl.*, vol. 15 (1967), p. 1.
23 Troster, H., and Brambring, M., *Child Care Health Dev.*, vol. 18 (1992), p. 207.
24 Marler, P., in Bateson, P. (ed.), *The Development and Integration of Behaviour* (Cambridge University Press, 1991), p. 41.
25 Sherry, D.F., and Galef, B.G., *Anim. Behav.*, vol. 40 (1990), p. 987.
26 Thayer, G.H., *Concealing-Coloration in the Animal Kingdom* (Macmillan, 1909).

6 *Alternative Lives*

1 Rowell, C.H.F., *Adv. Ins. Physiol.*, vol. 8 (1971), p. 145.
2 Janzen, F.J., and Paukstis, G.L., *Q. Rev. Biol.*, vol. 66 (1991), p. 149.
3 Yntema, C.L., and Mrosovsky, N., *Can. J. Zool.*, vol. 60 (1982), p. 1012.
4 Crews, D., *Zool. Sci.*, vol. 13 (1996), p. 1.
5 Wilson, E.O., *The Insect Societies* (Harvard University Press, 1971).
6 Pener, M.P., and Yerushalmi, Y., *J. Ins. Physiol.*, vol. 44 (1998), p. 365.
7 Lee, T.M., and Zucker, I., *Am. J. Physiol.*, vol. 255 (1988), R831.
8 Dunbar, R.I.M., *Reproductive Decisions* (Princeton University Press, 1984).
9 Pinker, S., *The Language Instinct* (Penguin, 1994).
10 Meck, W.H., and Williams, C., *Neuroreport*, vol. 8 (1997), p. 3045.
11 Barker, D.J.P., *Mothers, Babies and Health in Later Life* (Churchill Livingstone, 1998).
12 Godfrey, K.M., et al., *Br. Med. J.*, vol. 307 (1993), p. 405.
13 Hales, C.N., et al., *Diabetic Med.*, vol. 14 (1997), p. 189.
14 Waterlow, J.C., in Harrison, G.A., and Waterlow, J.C. (eds.), *Diet and Disease in Traditional and Developing Countries* (Cambridge University Press, 1990), p. 5.
15 Martin, P., *The Sickening Mind* (HarperCollins, 1997).
16 Ravelli, A.C.J., et al., *Lancet*, vol. 351 (1998), p. 173.

17 Stanner, S.A., et al., *Br. Med. J.*, vol. 315 (1997), p. 1342.
18 Artaud-Wild, S.M., et al., *Circulation*, vol. 88 (1993), p. 2771.
19 Parodi, P.W., *Med. Hypotheses*, vol. 49 (1997), p. 313.
20 Law, M., et al., *Br. Med. J.*, vol. 318 (1999), p. 1477.
21 Eveleth, P.B., and Tanner, J.M., *Worldwide Variation in Human Growth* (Cambridge University Press, 1990).
22 Huh, D.L., et al., *Int. J. Epidemiol.*, vol. 20 (1991), p. 1001.
23 Clutton-Brock, T.H., *The Evolution of Parental Care* (Princeton University Press, 1991).
24 Simmons, R., et al., *Can. J. Zool.*, vol. 64 (1986), p. 2447.
25 Alexander, R.D., et al., in Chagnon, N.A., and Irons, W. (eds.), *Evolutionary Biology and Human Social Behaviour* (Duxbury, 1979), p. 402.
26 Hauspie, R.C., et al., *Acta Paediat. Suppl.*, vol. 423 (1997), p. 20.
27 de la Puente, M.L., et al., *Ann. Hum. Biol.*, vol. 24 (1997), p. 435.

7 *Chance and Choice*

1 Barlow, N. (ed), *The Autobiography of Charles Darwin* (Collins, 1958).
2 Levy-Shiff, R., et al., *J. Pediat. Psychol.*, vol. 19 (1994), p. 63.
3 Wilkinson, R.G., *Unhealthy Societies* (Routledge, 1996).
4 Kidson, M.A., et al., *Med. J. Aust.*, vol. 158 (1993), p. 563.
5 Bremner, J.D., et al., *Am. J. Psychiat.*, vol. 153 (1996), p. 369.
6 North, C.S., and Smith, E.M., *Compr. Ther.*, vol. 16 (1990), p. 3.
7 D'Souza, D., *Br. J. Clin. Pract.*, vol. 49 (1995), p. 309.
8 Macksoud, M.S., and Aber, J.L., *Child Dev.*, vol. 67 (1996), p. 70.
9 Basoglu, M., et al., *J. Am. Med. Assoc.*, vol. 272 (1994), p. 357.
10 Pynoos, R.S., et al., *Br. J. Psychiat.*, vol. 163 (1993), p. 239.
11 Poulton, R.G., and Andrews, G., *Acta Psychiat. Scand.*, vol. 85 (1992), p. 35.
12 Jahoda, G., *Br. J. Psychol.*, vol. 45 (1954), p. 192.
13 Hofstadter, D.R., and Dennett, D.C., *The Mind's I* (Harvester, 1981).
14 Murphy, S.T., et al., *J. Pers. Soc. Psychol.*, vol. 69 (1995), p. 589.
15 Damasio, A.R., *Descartes' Error* (Grosset/Putnam, 1994).
16 Bechara, A., et al., *Science*, vol. 275 (1997), p. 1293.
17 Brickman, P., et al., *J. Pers. Soc. Psychol.*, vol. 36 (1978), p. 917.

8 Sensitive Periods

1 Bagheri, M.M., et al., *J. Perinat. Med.*, vol. 26 (1998), p. 263.
2 Sampson, P.D., et al., *Teratology*, vol. 56 (1997), p. 317.
3 Day, J.C., et al., *J. Neurosci.*, vol. 18 (1998), p. 1886.
4 Clarke, A.S., et al., *Inf. Behav. Dev.*, vol. 19 (1996), p. 451.
5 Dorner, G., et al., *Exp. Clin. Endocrinol.*, vol. 98 (1991), p. 141.
6 Rook, G.A., and Stanford, J.L., *Immunol. Today*, vol. 19 (1998), p. 113.
7 Howie, P.W., et al., *Br. Med. J.*, vol. 300 (1990), p. 11.
8 Wilson, A.C., et al., *Br. Med. J.*, vol. 316 (1998), p. 21.
9 Gordon, N., *Brain Dev.*, vol. 19 (1997), p. 165.
10 Fujinaga, T., et al., *Genet. Soc. Gen. Psychol. Monogr.*, vol. 116 (1990), p. 37.
11 Kuhl, P.K., et al., *Science*, vol. 255 (1992), p. 606.
12 Kuhl, P.K., et al., *Science*, vol. 277 (1997), p. 684.
13 Johnson, J.S., and Newport, E.L., *Cognition*, vol. 39 (1991), p. 215.
14 Acocella, J., *New Yorker* (19 January 1998), p. 45.
15 Snow, C., in Bornstein, M.H. (ed.), *Sensitive Periods in Development* (Erlbaum, 1987), p. 183.
16 Kim, K.S.H., et al., *Nature*, vol. 388 (1997), p. 171.
17 Marler, P., *J. Neurobiol.*, vol. 33 (1997), p. 501.
18 Bottjer, S.W., and Arnold, A.P., *Ann. Rev. Neurosci.*, vol. 20 (1997), p. 459.
19 Woodhead, J.A., et al., *Famine Inquiry Commission: Report on Bengal* (Government of India Publications Department, Department of Food, 1945).
20 Campos, E., *Surv. Ophthalmol.*, vol. 40 (1995), p. 23.
21 Freeman, T.C.B., et al., *J. Physiol.*, vol. 494P (1996), p. 18.
22 Liu, Y.L., et al., *Dev. Brain Res.*, vol. 79 (1994), p. 63.
23 Cohen, L.G., et al., *Nature*, vol. 389 (1997), p. 180.
24 Bolhuis, J.J., and Honey, R.C., *Trends Neurosci.*, vol. 21 (1998), p. 306.
25 Bateson, P., and Hinde, R.A., in Bornstein, M.H. (ed.), *Sensitive Periods in Development* (Erlbaum, 1987), p. 19.
26 Horn, G., *Memory, Imprinting and the Brain* (Clarendon Press, 1985).
27 Horn, G., et al., *Brain Res.*, vol. 56 (1973), p. 227.
28 Bateson, P.P.G., et al., *Brain Res.*, vol. 84 (1975), p. 207.

29 Bateson, P.P.G., et al., *Science*, vol. 181 (1973), p. 576.
30 Horn, G., *Trends Neurosci.*, vol. 21 (1998), p. 300.

9 *Morning Shows the Day*

1 Kagan, J., *Int. J. Behav. Dev.*, vol. 3 (1980), p. 231.
2 Hinde, R.A., and Bateson, P., *Int. J. Behav. Dev.*, vol. 7 (1984), p. 129.
3 Strohl, K.P., and Thomas, A.J., *Respir. Physiol.*, vol. 110 (1997), p. 269.
4 Perris, H., et al., *Child Dev.*, vol. 61 (1990), p. 1796.
5 Hepper, P.G., *Acta Paediat.*, vol. 85, (1996), p. 16.
6 Kagan, J., *Child Dev.*, vol. 68 (1997), p. 139.
7 Rutter, M., and Rutter, M., *Developing Minds* (Penguin, 1993).
8 Bowlby, J., in Bateson, P. (ed.), *The Development and Integration of Behaviour* (Cambridge University Press, 1991), p. 301.
9 Waters, E., et al., *Monogr. Soc. Res. Child Dev.*, vol. 60 (1995), p. 1.
10 Higley, J.D., et al., *Arch. Gen Psychiat.*, vol. 50 (1993), p. 615.
11 Suomi, S.J., *Br. Med. Bull.*, vol. 53 (1997), p. 170.
12 Benoit, D., and Parker, K.C.H., *Child Dev.*, vol. 65 (1994), p. 1444.
13 Werner, E.E., *Acta Paediat. Suppl.*, vol. 442 (1997), p. 103.
14 Benoit, T.C., et al., *Arch. Pediat. Adol. Med.*, vol. 150 (1996), p. 1278.
15 Lash, J.P., *Helen and Teacher* (Allen Lane, 1980).
16 Burger, J., and Gochfeld, M., *Pharmacol. Biochem. Behav.*, vol. 55 (1996), p. 339.
17 Klopfer, P., and Klopfer, M., *Anim. Behav.*, vol. 25 (1977), p. 286.
18 McCance, R.A., *Lancet*, vol. 2 (1962), p. 671.
19 Waddington, C.H., *The Strategy of the Genes* (Allen & Unwin, 1957).
20 Kauffman, S.A., *The Origins of Order* (Oxford University Press, 1993).
21 Damasio, A.R., *Descartes' Error* (Grosset/Putnam, 1994).
22 Goldman, P.S., and Galkin, T.W., *Brain Res.*, vol. 152 (1978), p. 451.
23 Lewin, R., *Science*, vol. 210 (1980), p. 1232.

24 Hidalgo, A., and Brand, A.H., *Development*, vol. 124 (1997), p. 3253.

10 Room 101

1 Pendergast, M., *Victims of Memory* (HarperCollins, 1996).
2 Bauer, P.J., et al., *Memory*, vol. 2 (1994), p. 353.
3 Hinde, R.A., et al., *Dev. Psychol.*, vol. 21 (1985), p. 130.
4 Shapiro, D.Y., *J. Exp. Zool.*, vol. 261 (1992), p. 194.
5 Sulloway, F.J., *Born to Rebel* (Little, Brown, 1996).
6 Watson, P., *War on the Mind* (Hutchinson, 1978).
7 Sargant, W., *Battle for the Mind* (Heinemann, 1957).
8 Rowe, M.K., and Craske, M.G., *Behav. Res. Ther.*, vol. 36 (1998), p. 719.
9 Gellner, E., in Fabian, A.C. (ed.), *Origins* (Cambridge University Press, 1988), p. 128.
10 Graham, D.L., et al., *Violence Vict.*, vol. 10 (1995), p. 3.
11 Bateson, P., in Oliverio, A., and Zappella, M. (eds.), *The Behavior of Human Infants* (Plenum Press, 1983), p. 57.
12 Martin, P., *The Sickening Mind* (HarperCollins, 1997).
13 Pettigrew, J.D., and Kasamutsu, T., *Nature*, vol. 271 (1978), p. 761.
14 Insel, T.R., et al., *Rev. Reprod.*, vol. 2 (1997), p. 28.
15 Young, L.J., et al., *Trends Neurosci.*, vol. 21 (1998), p. 71.
16 Leckman, J.F. et al., *Psychoneuroendocrinol.*, vol. 19 (1994), p. 723.

11 Everything to Play For

1 Darwin, C., *Mind*, vol. 2 (1877), p. 285.
2 Deci, E.L., and Ryan, R.M., *Adv. Exp. Soc. Psychol.*, vol. 13 (1980), p. 39.
3 Martin, P., and Caro, T.M., *Adv. Study Behav.*, vol. 15 (1985), p. 59.
4 Fagen, R., *Animal Play Behavior* (Oxford University Press, 1981).
5 Lee, P.C., *Behaviour*, vol. 91 (1984), p. 245.
6 Harcourt, R., *Anim. Behav.*, vol. 42 (1991), p. 509.
7 Boden, M.A., *Piaget* (Harvester Press, 1979).
8 Gomendio, M., *Anim. Behav.*, vol. 36 (1988), p. 825.
9 Caro, T.M. *Anim. Behav.*, vol. 49 (1995), p. 333.

10 Stamps, J., *Am. Nat.*, vol. 146 (1995), p. 41.
11 Einon, D., and Potegal, M., *Aggress. Behav.*, vol. 17 (1991), p. 27.
12 Furedi, F., *Culture of Fear* (Cassell, 1997).

12 Sex, Beauty and Incest

1 Buss, D.M., *The Evolution of Desire* (Basic Books, 1994).
2 Perrett, D.I., et al., *Nature*, vol. 368 (1994), p. 239.
3 Singh, D., *J. Pers. Soc. Psychol.*, vol. 65 (1993), p. 293.
4 Kowner, R., *Genet. Soc. Gen. Psychol. Monogr.*, vol. 122 (1996), p. 215.
5 Jennions, M.D., and Petrie, M., *Biol. Rev.*, vol. 72 (1997), p. 283.
6 Darwin, C.R., *The Descent of Man, and Selection in Relation to Sex* (John Murray, 1871).
7 Andersson, M., *Sexual Selection* (Princeton University Press, 1994).
8 Zahavi, A., and Zahavi, A., *The Handicap Principle* (Oxford University Press, 1997).
9 ten Cate, C., and Bateson, P., *Evolution*, vol. 42 (1988), p. 1355.
10 Miller, G.F., *Evol. Hum. Behav.*, vol. 19 (1988), p. 343.
11 Bruce, V., and Young, A., *In the Eye of the Beholder* (Oxford University Press, 1998).
12 Møller, A., *Nature*, vol. 332 (1988), p. 640.
13 Møller, A.P., and Thornhill, R., *Am. Nat.*, vol. 151 (1998), p. 174.
14 Bateson, P., in Bateson, P. (ed.), *Mate Choice* (Cambridge University Press, 1983), p. 257.
15 Shepher, J., *Arch. Sex. Behav.*, vol. 1 (1971), p. 293.
16 Wolf, A.P., *Sexual Attraction and Childhood Association* (Stanford University Press, 1994).
17 Hilson, S., *Teeth* (Cambridge University Press, 1986).
18 Brown, J., and Eklund, A., *Am. Nat.*, vol. 143 (1994), p. 435.
19 Wedekind, C., et al., *Proc. Roy. Soc. Lond. B*, vol. 260 (1995), p. 435.
20 Wedekind, C., and Furi, S., *Proc. Roy. Soc. Lond. B*, vol. 264 (1997), p. 1471.
21 Tooby, J., and Cosmides, L., in Barkow, J.H., et al. (eds.), *The Adapted Mind* (Oxford University Press, 1992), p. 19.
22 Alexander, R.D., et al., in Chagnon, N.A., and Irons, W. (eds.),

Evolutionary Biology and Human Social Behavior (Duxbury Press, 1979), p. 402.

23 Langlois, J.H., et al., *Dev. Psychol.*, vol. 26 (1990), p. 153.

24 Walton, G.E., and Bower, T.G.R., *Psychol. Sci.*, vol. 4 (1993), p. 203.

25 Maccoby, E.E., *Am. Psychol.*, vol. 45 (1990), p. 513.

26 Sabol, S.Z., et al., *Hum. Genet.*, vol. 103 (1998), p. 273.

27 Bailey, J.M., and Pillard, R.C., *Arch. Gen. Psychiat.*, vol. 48 (1991), p. 1089.

28 Blanchard, R., and Bogaert, A.F., *Am. J. Psychiat.*, vol. 153 (1996), p. 27.

29 Blanchard, R., and Klassen, P., *J. Theor. Biol.*, vol. 185 (1997), p. 373.

30 Hall, J.A., and Kimura, D., *Behav. Neurosci.*, vol. 108 (1994), p. 1203.

31 Swaab, D.F., et al., *J. Homosex.*, vol. 28 (1995), p. 283.

32 Dorner, G., et al., *Exp. Clin. Endocrinol.*, vol. 98 (1991), p. 141.

33 LeVay, S., *Science*, vol. 253 (1991), p. 1034.

34 Cantoni, D., and Brown, R.E., *Anim. Behav.*, vol. 54 (1997), p. 377.

35 Morris, F.O., *A History of British Birds* (Groombridge & Sons, 1853).

36 Davies, N.B., *Dunnock Behaviour and Social Evolution* (Oxford University Press, 1995).

37 Thornhill, N.W., *Behav. Brain Sci.*, vol. 14 (1991), p. 247.

38 Westermarck, E., *The History of Human Marriage* (Macmillan, 1891).

13 This Strange Eventful History

1 Bowlby, J., *Attachment and Loss*, vol. 1, *Attachment* (Hogarth Press, 1969).

2 Kessen, W., *The Rise and Fall of Development* (Clark University Press, 1990).

3 Groebel, J., and Hinde, R.A., *Aggression and War* (Cambridge University Press, 1989).

4 Kauffman, S.A., *The Origins of Order* (Oxford University Press, 1993).

Further Reading

Manning and Dawkins give an excellent introduction to the study of animal behaviour.[1] Michel and Moore provide a more advanced approach to the biology and psychology of behavioural development.[2] The view that inheritance involves both the genes and the environment in which the individual develops is elegantly presented by Oyama.[3] Her systems approach is brought up to date by many of the contributors to the volume edited by Lerner.[4] An authoritative view of human development across the whole lifespan is given by the Rutters.[5] It is also provided by experts in many fields in the book edited by Magnusson.[6] With wonderfully described examples, Dawkins shows how the idea of natural design may be placed in the context of modern Darwinism.[7] Coen describes the revolution in the understanding of how genes are involved in the creation of an adult from a fertilised egg.[8] Stewart and Cohen present an entertaining account of how complex behaviour evolved and how it develops in the individual.[9] Rose argues eloquently that the uniqueness of the individual has to be understood in terms of both the biology and the events that create what he calls the 'lifeline'.[10] Hinde gives a masterly survey of the role of personal relationships in human development.[11]

1. Manning, A. and Dawkins, M.S., *An Introduction to Animal Behaviour* 5th edition (Cambridge University Press, 1998).

2. Michel, G.F. and Moore, C.L., *Developmental Psychobiology* (MIT Press, 1995).
3. Oyama, S., *The Ontogeny of Information* (Cambridge University Press, 1985).
4. Lerner, R.M. (ed.), *Handbook of Child Psychology. Vol. 1* 5th edition (Wiley, 1998).
5. Rutter, M. and Rutter, M., *Developing Minds* (Penguin, 1993).
6. Magnusson, D. (ed.), *The Lifespan Development of Individuals* (Cambridge University Press, 1995).
7. Dawkins, R., *The Blind Watchmaker* (Longman, 1986).
8. Coen, E., *The Art of Genes* (Oxford University Press, 1999).
9. Stewart, I. and Cohen, J., *Figments of Reality* (Cambridge University Press, 1997).
10. Rose, S., *Lifelines* (Allen Lane, 1997).
11. Hinde, R.A., *Relationships* (Psychology Press, 1997).

Index

Index

Index

Index